Infectious Disease: A Very Short Introduction

VERY SHORT INTRODUCTIONS are for anyone wanting a stimulating and accessible way into a new subject. They are written by experts, and have been translated into more than 45 different languages.

The series began in 1995, and now covers a wide variety of topics in every discipline. The VSI library currently contains over 700 volumes—a Very Short Introduction to everything from Psychology and Philosophy of Science to American History and Relativity—and continues to grow in every subject area.

Very Short Introductions available now:

Available soon:

For more information visit our website

www.oup.com/vsi/

Marta L. Wayne and Benjamin M. Bolker

INFECTIOUS DISEASE

A Very Short Introduction

SECOND EDITION

OXFORD
UNIVERSITY PRESS

Great Clarendon Street, Oxford, OX2 6DP,
United Kingdom

Oxford University Press is a department of the University of Oxford.
It furthers the University's objective of excellence in research, scholarship,
and education by publishing worldwide. Oxford is a registered trade mark of
Oxford University Press in the UK and in certain other countries

© Marta L. Wayne & Benjamin M. Bolker 2023

The moral rights of the authors have been asserted

First edition published 2015

Published in the United States of America by Oxford University Press
198 Madison Avenue, New York, NY 10016, United States of America

British Library Cataloguing in Publication Data
Data available

Library of Congress Control Number: 2023935107

ISBN 978-0-19-285851-1

Printed and bound by
CPI Group (UK) Ltd, Croydon, CR0 4YY

Contents

Acknowledgements

For Charlie, Norma, Tara, and (*in memoriam*) Django. We acknowledge the partial support of the US National Institutes of Health. We would like to thank the understanding and professional editorial staff at Oxford University Press, especially Cathy Kennedy, along with our colleagues who contributed friendly reviews and technical advice: Rustom Antia, Janis Antonovics, Ottar Bjørnstad, Julia Buck, Robin Bush, Derek Cummings, Jonathan Dushoff, David Earn, Tom Hladish, David Hillis, Lindsay Keegan, Marm Kilpatrick, Aaron King, Marc Lipsitch, Ana Longo, Glenn Morris, Juliet Pulliam, Marco Salemi, and David Smith. Needless to say, we are entirely responsible for any remaining errors or oversimplifications.

List of illustrations

Chapter 1
Infection is inevitable

We talk about 'infectious lyrics' and 'viral videos'. Metaphors of
infectious disease saturate Western popular culture in the 21st
century. The COVID-19 pandemic of the early 2020s has focused
the world's attention on infectious disease, but it is only the latest
infectious disease of the 21st century (after SARS, H1N1 influenza,
Ebola, and Zika) and it will not be the last.

As recently as the 1970s, doctors were boldly proclaiming the
beginning of the end for infectious disease. They thought their
arsenal of vaccines for preventing viral diseases and broad-spectrum
antibiotics for treating bacterial infections could handle any threat.
But disease was never dead, or even in remission. Even as the
doctors announced victory, drug resistant strains of *Staphylococcus
aureus* (one of the 'flesh-eating bacteria' of British tabloids) were
spreading in hospitals. (Japan had experienced outbreaks of drug
resistant bacteria in the 1950s, but at the time these epidemics
were little noticed in the West.) Things got worse as HIV, which
has stubbornly resisted the development of vaccines to the present
day, emerged in the 1980s. In recognition of the renewed threat of
infectious disease, the US Institute of Medicine coined the phrase
'emerging and re-emerging diseases' in the early 1990s.

An infectious disease is one that you can catch from another
person or organism as a result of the transmission of a biological

agent. In contrast, you fall ill with non-infectious diseases—heart disease, diabetes, Alzheimer's—because of a combination of your environment and genes inherited from your parents. The agents that cause infectious disease are called *pathogens* or (more broadly) *parasites*. While biologists once used 'parasite' only to describe relatively large disease-causing agents such as tapeworms or ticks, they now include microorganisms (viruses, bacteria, fungi, and protists) in this category as well, making the two terms more or less synonymous.

Infectious disease frightens us precisely *because* it is infectious. Its agents are invisible to the naked eye and thus largely unavoidable, except by entirely eschewing human contact. Edgar Allan Poe illustrated the fear of infection, as well as the futility of cutting off human contact to evade infectious disease, in his 1842 story *The Masque of the Red Death*. In Poe's story, a group of wealthy nobles withdraw to an isolated location to escape from a plague called the Red Death. Ultimately, a costumed stranger infiltrates the group at a masquerade ball. Despite their precautions, the entire group succumbs to the disease.

However ineffective it might be, for most of human history the strategy taken by Poe's nobles—avoiding disease transmission—has been the only way to combat infectious disease. As the COVID-19 pandemic has shown, this strategy remains necessary even in the 21st century. Since the mechanisms of disease transmission were unknown until the mid-19th century, all human societies could do in the face of an epidemic was to cut off contact with infected areas. In 1665, Isaac Newton retreated to the countryside to avoid the Great Plague of London—and incidentally invented calculus and discovered the law of gravity. In the same year, the English village of Eyam voluntarily quarantined itself to prevent the spread of the plague, with half or more of the villagers ultimately dying. The word 'quarantine', which now describes the compulsory isolation of potentially infected people to avoid transmission to others, is derived from 'quaranta giorni', the 40 days that ships had to wait outside the city of Venice to be sure they were free of plague.

Quarantines (at least those better than the one in Poe's story) do block transmission, but they are fundamentally reactive—they are only imposed once we become aware of a serious threat of disease. They help only healthy people living in uninfected populations, not individuals who have already been infected or uninfected people unlucky enough to be stuck in the quarantine zone. On the other hand, quarantines can be effective against any disease, provided that we know something about its mode of transmission (since plague is generally spread by rat fleas, preventing communication among humans while allowing rats to move freely is useless).

Quarantines are deployed to protect groups of people, rather than individuals. As medical science improved, public health officials began to shift their focus from the protection of populations to the protection of individuals. Immunization—the process of protecting people by stimulating their immune systems with foreign substances such as mild pathogen strains or toxins—was the first of several major breakthroughs in individual-focused infectious disease control. Immunization for smallpox was widely practised in Africa, China, India, and Turkey by the early 18th century. It achieved public visibility in the West following its importation to England in 1721 by Lady Mary Wortley Montagu, the wife of the British ambassador to Turkey, and more spectacularly via an 'experiment' in the same year promoted by clergyman Cotton Mather of the colonial city of Boston, Massachusetts. In the face of a smallpox outbreak, Mather and his medical colleague Zabdiel Boylston promoted immunization rather recklessly, against the will of the majority of his fellow Bostonians. Despite several deaths, the experiment demonstrated an effective alternative to physical isolation: immunization protected individuals from infection without restricting anyone's freedom of movement.

Mather and Boylston's experiment also illustrated an ethical conflict between controlling disease for the benefit of an individual and controlling disease for the benefit of an entire population. Bostonians who were successfully immunized were safe from disease, but for several days following immunization they could have transmitted the disease to unprotected individuals. Most modern immunizations involve non-infectious substances, so this particular problem is of lesser concern today, but the conflict between individual and public health, and between individual rights and public health, is very much alive.

Most immunizations can only prevent healthy individuals from becoming infected, or reduce the impact of infection, not cure infected people. Individual-level control of disease took another leap forward in the mid-20th century with the advent of antibiotic chemicals. First derived from common household moulds, bacteria, and even fabric dyes, antibiotics could be used to cure individuals who were already infected. The possibility of curing disease also lessened the fear of quarantine, which had previously been seen as a death sentence.

Between antibiotics to cure harmful bacterial infections and a wave of new and effective vaccines to prevent diseases such as polio, measles, and pertussis, an infectious disease-free future must have seemed within reach to the public health officials of the 1970s. However, public health officials were quickly faced with proof that individual-level control fails for many diseases. Vaccines work by priming the human immune system, and thus they are much harder to develop for disease agents such as malaria or HIV that have evolved strategies for evading the immune system. Antibiotics are only effective against bacteria, not other microorganisms such as viruses or fungi (while antiviral and antifungal chemicals do exist, they are much less broadly effective than antibiotics). With the realization in the late 20th century that infectious diseases were not vanquished after all, research began to shift back towards population-level control.

So far we have divided treatments according to whether they primarily help populations (quarantine) or individuals (immunization/vaccination, antibiotics). Looking more closely, however, we can see that both antibiotics and immunization do help protect populations, as well as benefiting the individuals who receive treatment. Using drugs to cure sick people reduces the impact of infectious disease, because people who recover also stop infecting others. Thus, treating sick people can reduce transmission. Using vaccines to protect people from infection means that some potentially infectious contacts (activities by infected people such as sneezing or sexual activity, depending on the disease) are wasted on people who are protected from disease, again reducing transmission. This so-called *herd immunity* reduces the size of an epidemic even beyond the direct effects of vaccination. If we immunize enough of the population, we can reduce transmission sufficiently to stamp out an epidemic. If we can do this at a global scale, then the disease will become extinct (as in the case of smallpox, one extinction that doesn't bother environmentalists).

If the problem were just that some diseases are harder to control than others, we would still be making progress, albeit slowly, in the fight against infectious disease. Modern molecular biology has provided us with a variety of new antiviral drugs, and vaccines are in development even for such difficult cases as malaria and HIV. But both humans and infectious disease agents are living organisms, and all living organisms undergo ecological and evolutionary change, making infectious disease a moving target. Our growing recognition that we (and our plagues!) are tied to the wheel of life, and our realization that individual-level approaches have failed to free us from the wheel, drives the shift in infectious disease research today.

Ecological processes

As much as we try to deny it, humans are subject to the laws of ecology. We control most aspects of our environment. Motor vehicles have replaced large predators as the leading category of violent death; we have wiped out most of our potential competitors; and we have domesticated the organisms below us in the food chain. But infectious disease still connects us to ecology's global web.

The most important disease–ecology connection is *zoonosis*, the transmission of new diseases from animal *reservoir hosts* to humans: COVID-19 may be the most spectacular example, but many of the new and emerging diseases that we haven't got a handle on come from animals: Ebola, SARS, avian (H5N1) influenza, and hantavirus are a few of the better-known examples. In fact, almost all infectious diseases originate in this way, and most emerging disease threats are zoonotic. Since it is difficult to vaccinate or design drugs against unknown diseases, this parade of new threats is terrifying: we don't know when the next 'super-disease' might emerge.

Zoonoses are as old as humanity. Smallpox is thought to have moved from rodents into humans at least 16,000 years ago; measles probably moved from cattle to humans when humans first started to live in large cities; and HIV jumped from monkeys and chimpanzees to humans in the early 20th century. However, rapid human population growth and changes in land use have increased human–animal contact, whether in the tropical rainforest (HIV and Ebola) or in the temperate suburbs (Lyme disease).

As well as coming into more contact with animals, humans are moving around the planet at an ever-increasing pace. Contact between individuals, and thus transmission, can happen much faster and over much greater distances than when Isaac Newton moved to the countryside to escape the plague in the 17th century or when Poe penned *The Masque of the Red Death* in the 19th century. Diseases that had previously been confined to narrow regions (generally low income countries) can rapidly expand their ranges. This is true not only for human infectious diseases, but also for diseases affecting other species whose infectious agents are transported by humans in our luggage, in the food with which we sustain ourselves during travel, or on our shoes. Human travel and commerce can spread disease indirectly, by transporting *vectors*: animals (especially insects) that transmit disease from one organism to another. For example, the international trade in used tyres is spreading *Aedes* mosquitoes, the vector of dengue fever. As well as vectors, we sometimes move the reservoir hosts of zoonoses. The first human infections of the Ebola virus outside Africa, in 1989, came from monkeys (crab-eating macaques) that had been imported from the Philippines for animal experimentation: luckily, the particular strain involved (Ebola Reston) turned out to be harmless to humans.

Increasing movement spreads vectors and hosts to new areas; environmental change allows them to thrive in their new homes. With global climate change, animals and especially temperature-sensitive insects can invade new areas in temperate regions. Although the topic is still controversial, many climate scientists and some epidemiologists are convinced that mosquito-borne diseases like dengue and malaria are already spreading to new populations under the influence of regional climate change. An even greater impact comes from more localized environmental changes driven by human patterns of settlement and economic activity. For example, the larvae of dengue-transmitting mosquitoes thrive in water bodies as small as used tyres and household water tanks. More generally, as people move from rural to ever-growing urban environments, they face greater sewage problems (spreading cholera and other water-borne disease) and encounter new and different kinds of disease-bearing insects.

Evolutionary processes

Ecology constantly exposes us to new epidemics, but evolution is even worse: the diseases we already know change even as we attempt to come to grips with them. As living organisms fighting for survival, infectious diseases don't accidentally escape our attempts to control them. They are groomed by natural selection to escape. Infectious disease is a moving target that moves faster the harder we try to hit it. Disease biologists frequently invoke Lewis Carroll's Red Queen from *Through the Looking Glass*, who said: 'it takes all the running you can do, to keep in the same place'.

For every disease prevention strategy, infectious diseases have an evolutionary countermeasure. Bacteria did not evolve antibiotic resistance in response to human antibiotic use: scientists have found antibiotic resistance genes similar to modern variants in DNA extracted from 30,000-year-old frozen soil. This isn't surprising, because humans did not invent most antibiotics. Rather, we borrowed or co-opted them from fungi, which had evolved them as a strategy for combating bacteria. However, the widespread use of antibiotics in both medicine and agriculture has allowed bacteria that are resistant to one or more types of antibiotic to outcompete their susceptible counterparts. Other organisms, such as the protozoans that cause malaria, have also evolved resistance to the drugs used to treat them. And when HIV patients are given a single drug rather than a multi-drug 'cocktail', the virus evolves drug resistance within their bodies in just a few weeks. Pathogens evolve resistance to vaccines as well as drugs, but in a different way. Rather than resistance genes spreading within the pathogen population, *strain replacement* occurs— previously rare types that are immune to our vaccines, such as the Omega strain of SARS-CoV-2, take over the population.

Although mosquitoes have smaller populations and lower birth rates than bacteria and viruses, and hence evolve much more slowly, they too have found evolutionary countermeasures to our disease control strategies. In high income countries, DDT use was discontinued as Rachel Carson and others spread the alarm about its harmful effects on wildlife, but vector control strategies based on DDT were short-lived even in middle and low income countries because DDT resistant mosquitoes evolved within a decade of the onset of mass spraying programmes.

Every aspect of infectious disease biology, not just the ability to resist or circumvent control measures, is constantly evolving. Biologists have noted that the virulence of a disease—how harmful it is to its host—is an evolutionary characteristic of the pathogenic organism. Typically mild diseases can suddenly acquire mutations that make them much nastier. A small number of mutations in the West Nile virus (WNV) that arose in the late 1990s made it far more lethal to birds and mice (and probably humans, although it's hard to know for sure since we don't experiment on humans).

Although mutations are random, evolution by natural selection is not: once pathogens mutate, their subsequent success depends on ecological conditions. Biologist Paul Ewald was among the first to point out that changes in pathogens' ecological conditions, such as a shift from direct person-to-person transmission to water-borne transmission, could favour more virulent forms of infectious diseases. The rise of global air travel may drive evolutionary as well as ecological changes in disease: some biologists have pointed out that mixing between spatially separated populations can encourage virulence, although so far there are no verified real-world examples of this phenomenon.

Outlook

Given these challenges, the elimination of infectious disease—the siren song of the 20th century—seems hopelessly naive, and approaches based solely on protecting individuals appear both untenable and unjust, given inequality in access to healthcare. It would seem that we must learn to live with infectious disease, rather than eliminate it. However, we must also strive to reduce the misery caused by infectious disease. Accordingly, this century has seen a shift from attempts to eliminate the agents of infectious disease, to attempts to understand, predict, and manage infectious disease transmission at the population level. Synergistic approaches, not simply magic bullets, are required for a sustainable approach for everyone to be able to live well with infectious disease: we must make use of tools from molecular biology, economics, and sociology, among others. We will touch on many of these topics, but focus our book on the disciplines of ecology and evolution: ecology, because understanding ecological relationships helps us understand cycles of transmission; evolution, because disease agents evolve, both on their own and in response to our efforts to control them.

Chapter 2
Transmission at different scales

Transmission defines infectious disease. Transmission occurs when someone passes a disease to someone else: technically speaking, when a pathogen that was established in one host organism's body succeeds in moving into another host's body and establishing itself there.

Transmission occurs in a huge variety of ways. For example, in transmission of respiratory diseases such as influenza, virus particles produced by the cells in an infected person's lungs are first coughed or sneezed into the surrounding atmosphere. These infectious particles can survive briefly in the air or on surfaces in the environment, and thus be directly transmitted from person to person with minimal contact. The receiving person can either inhale them directly, or can pick them up by touching a surface shortly after virus-containing droplets land there. The receiver can then transfer virus particles to their nose by touching their face; from there, the natural movement of air within their nose moves the virus into their respiratory tract. In the respiratory tract, the virus particles enter vulnerable cells and resume their cycle of spreading from one cell to another within the host's body.

Many viruses, including influenza and diarrhoea-causing viruses such as rotavirus, can survive for days in the environment, building up on particular kinds of objects known as *fomites*.

Fomites may even push male physicians to wear bowties: ever since health researchers identified regular neckties as potential fomites, arguments have raged in the medical community about fashion vs disease control. Influenza viruses can even survive for several days on banknotes, especially if they are first mixed with 'nasopharyngeal secretions' (snot), although we don't know how much these survivors contribute to disease transmission.

Pathogens whose infectious particles die very quickly outside the warm, wet environment of the human body often rely on bodily fluids being directly transferred from person to person, as in the case of sexually transmitted diseases such as HIV (see Chapter 4) and gonorrhoea. While sexual contact was the most common form of fluid exchange throughout most of human evolutionary history, these pathogens can also be transmitted by more modern modes of fluid exchange such as blood transfusions or the sharing of syringes by drug users.

Other pathogens that cannot survive in the environment have evolved to use biological organisms, especially blood-sucking insects, ticks, and mites, as *vectors* to travel from one host to another. This strategy requires considerably more biological machinery than direct transfer between the bodies of two hosts of the same species. In the extreme case of pathogens with complex life cycles such as malaria (see Chapter 6), the pathogen goes through major transformations within the body of the mosquito vector. In fact, from the perspective of a mosquito-inhabiting malaria parasite, a human is just a convenient way to transmit itself to another mosquito.

Other infectious diseases can persist much better outside their hosts' bodies. The agents causing diseases such as cholera (see Chapter 5), typhoid, and Legionnaires' disease can survive in water, making their way from one host to another through drinking water or air conditioning systems. Anthrax—which kills its hosts quickly, reducing the potential for direct transmission from one animal host to another—produces long-lasting spores that survive for years in the environment, infecting grazing animals years later when they ingest spores attached to soil particles. Many fungi, such as certain species of *Aspergillus*, live primarily as free-living organisms, but can sometimes grow within human hosts if they find themselves there, especially if the host has its immune system weakened by stress or infection with other diseases. (In contrast to the *obligate* host dependence of most pathogens, such *opportunistic* pathogens can live in a host if one is available, but do not require a host in order to complete their life cycles.) The amphibian fungus *Batrachochytrium dendrobatidis* (Chapter 7) is closely related to non-pathogenic soil-dwelling fungi, but is itself an obligate parasite—as far as we know it can only persist in the environment for a few weeks.

Filters for encounter and compatibility

Following the work of Claude Combes, we can break down the process of transmission from one host to another into three stages: (1) transfer of infectious particles from inside the original host's body to the environment; (2) transfer of infectious particles through the environment, or through the bodies of intermediary vectors or hosts, to the receiving host; (3) transfer of particles from the environment into parts of the receiving host's body such as the blood, lungs, or liver where the pathogen can reproduce. These three stages collectively comprise the *encounter filter*.

Having made it into a new host's body, the travelling pathogen must overcome physical, biochemical, and immunological barriers in order to grow in the body of the new host. In other words, even if the pathogen can pass the encounter filter, it must also be biologically compatible with the new host; this final stage is called the *compatibility filter*. A host could close its compatibility filter by having a disease-resistant genetic mutation, such as the sickle-cell variant of the haemoglobin gene that protects against malaria, or the CCR5-Δ32 mutation that protects against HIV.

Opportunistic fungal infections are usually blocked as long as the host has a properly functioning immune system. In order to block most viral diseases, however, the host's immune system needs to have encountered the pathogen before, either naturally or through vaccination.

Both the encounter and compatibility filters must be open in order for successful transmission to occur. Public health measures can close the encounter filter and are especially important in the early stages of an epidemic. Drugs or vaccines can close the compatibility filter, but they are not always available.

Methods for closing the encounter filter include simple preventive strategies such as quarantine (see Chapter 1). They also include environmental strategies such as improved sanitation to control water-borne disease, or mosquito and tick control to stop vector-borne disease. Another class of strategies involves trying to convince people to modify their behaviour. These include all the rules that have become so well known during the COVID-19 pandemic (stay 2 metres away from other people, avoid indoor gatherings), or the US Centers for Disease Control's suggestions for avoiding mosquito-borne diseases such as West Nile virus: stay indoors at dusk, wear long pants and long-sleeved shirts, and use insect repellent. Though changing people's behaviour is difficult, it is sometimes the cheapest way to control disease. You don't need to inject or swallow substances that may have harmful side effects, and behavioural changes can even protect against unknown pathogens. Avoiding exchanging bodily fluids with strangers is a good idea, even if they have been screened for all currently known diseases.

We can rarely control disease with a single filter. The *Swiss cheese model*, first introduced in aviation safety, emphasizes that individual filters may fail or be unusable by some people, and that we often need to combine different kinds of controls at the environmental, population, and individual levels, using tools from ecology, behavioural psychology, and molecular biology. For malaria, we need bed-nets *and* indoor spraying *and* antimalarial drugs *and* vaccines; for Lyme disease, we need to cut brushy vegetation around our houses *and* control deer populations *and* check ourselves for ticks.

Epidemic dynamics

It's easy to understand the encounter and compatibility filters at the individual level: if you can prevent the transfer of infectious particles from the environment into your body, or if you immunize yourself to prevent the infection from taking hold in your body, you can stay safe. In order to understand the effects of these filters at the population level—for example, to decide whether an immunization programme or a quarantine will stop an epidemic—we need mathematical models. Almost as soon as biologists began to understand the mechanics of disease transmission, mathematicians started to develop models to describe the effects of the encounter and compatibility filters at the population level. As early as 1760, Daniel Bernoulli, a member of an eminent Swiss family of mathematicians and scientists, used a mathematical model to describe how smallpox immunization (i.e. closing the compatibility filter for some individuals) could improve public health. Bernoulli concluded that immunization could increase the expected lifespan at birth by 10 per cent, from about 27 to 30 years (the expected lifespan at birth was very short in the 18th century because of the high rate of infant and childhood mortality).

Bernoulli's model only took into account the direct benefits of immunization, thus missing the key insight of *herd immunity*. Immunization protects the people who are immunized, but it also reduces the prevalence of the disease and thus provides an indirect benefit to non-immunized people. To eradicate disease, you don't need to close the compatibility and encounter filters entirely (i.e. immunize 100 per cent of the people, or prevent transmission 100 per cent of the time); you just need to reduce transmission enough so that each infectious case gives rise to less than one new case. In technical terms, you need to reduce the *reproductive number*—the average number of new cases generated by a single case—to less than 1. If you succeed, then the disease will die out in the population as a whole, even though a few unlucky people may still get infected.

The reproductive number depends on the biology of the disease: how quickly can it produce new infectious particles? How well do they survive in the environment? It also depends on the ecology and behaviour of the host, which controls the encounter filter: how often do hosts run into each other, and how do they interact when they do? Are they washing their hands or wearing masks? Finally, it depends on the fraction of the population that remains susceptible to the disease, which declines over the course of an epidemic as individuals first get infected and then recover (typically becoming immune, at least temporarily) or die; as we have seen during the COVID-19 pandemic, behaviour also changes over the course of the epidemic as individuals' fear of diseases waxes and wanes. To ignore these last complications, epidemiologists focus on the *intrinsic reproductive number*, R_0 (pronounced 'R-zero' or 'R-nought'), which is the number of cases that would be generated by the first case in a new outbreak. R_0 is a basic measure of disease biology and community structure; it doesn't depend on how far the epidemic has spread through the population. If you can close the compatibility and encounter filters far enough to reduce the intrinsic reproductive number to less than 1, then you can not only control an epidemic in progress, but prevent the disease from getting started in the first place.

The importance of this kind of average-centred, population-level thinking in disease control was first appreciated by Ronald Ross, who built mathematical models of malaria transmission to prove that malaria could be eradicated without completely eliminating mosquitoes, by reducing mosquito populations below a threshold level—so that on average each infected human led to less than one new human case. (As we will see in Chapter 6, mosquito control and other methods for closing the encounter and compatibility filters have successfully eradicated malaria in some places, but not worldwide.) Ross won the Nobel Prize in 1902 for elucidating the life cycle of malaria, but his biography at the Nobel Foundation's website states that 'perhaps his greatest [contribution] was the development of mathematical models for the study of [malaria] epidemiology'.

Ross's model was one of the first *compartmental models*, which divide the population into compartments according to their disease status and track the rates at which individuals change from one disease status to another. The simplest compartmental model is called the *SIR model* because it divides the population up into *Susceptible*, *Infective*, and *Recovered* (or in some cases *Removed*) people. Susceptibles are people who could get infected, but are not currently infected (i.e. their compatibility filter is open); infectives have the disease and can transmit it (i.e. they are *infectious* as well as infected); recovered people have had the disease and are at least temporarily immune. ('Removed' is used for people who die from infection or animals that are killed to stop them from infecting others—while the difference between recovery and death matters to an individual, they have the same consequences for epidemic spread…)

The original compartmental models spawned many variations: for example, SIS models represent diseases such as gonorrhoea where individuals go straight back into the susceptible compartment once they have been cured of disease (say by taking antibiotics), because there is no effective immunity. Dozens of books and thousands of scientific papers have been written about compartmental models. Researchers have added all kinds of complexity to these models, accounting for the effects of genetics, age, and nutrition on the compatibility filter, and incorporating social and spatial networks to model the encounter filter.

Compartmental models also form the basic structure of huge *agent-based* computer models that track the behaviour and infection status of every individual in the population in order to understand the spread of epidemics such as influenza or COVID-19.

While realism and faithfulness to the biological facts of a given disease are important, compartmental models have remained the workhorse of epidemiological modelling because, even in their simpler forms, they capture most of the important characteristics of the spread of disease through a population. Especially when we are ignorant of important information about a disease—a situation painfully familiar to epidemiologists—an oversimplified model can be more useful than an overcomplicated one, as long as we interpret its conclusions cautiously.

Compartmental models typically assume that everyone in the population starts out equally susceptible to a particular disease (at or soon after birth, or in the case of sexually transmitted diseases, once they become sexually active). Susceptibles get infected by mixing with infected people in some way—for example, being coughed or sneezed on or exchanging bodily fluids. The infection rate increases with the proportion of infected people in the population. After an *infectious period* during which they spread disease, infected people recover; they move into the recovered compartment and gain effective immunity to the disease. A huge number of variations on this model are possible, including subdividing the population by age, sex, or geographic location; allowing people to return to the susceptible class from the recovered class after some time period; or allowing for variation in the rate at which different individuals transmit disease.

The structure of the SIR model (Figure 1) helps categorize the ways we can control epidemics. The most common control strategy—closing the compatibility filter by immunization or *prophylactic* drug treatment (i.e. giving people drugs to prevent

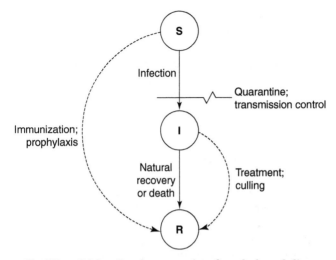

1. The SIR model describes the progression of people through disease stages from susceptible to infectious to recovered/removed. Interventions such as culling, treatment, or quarantine can speed up or prevent transitions between compartments.

rather than cure disease)—moves individuals directly from the susceptible to the recovered compartment without passing through the infected compartment on the way, at least until their immunity wanes or they stop taking the treatments. Most other epidemic control measures affect the encounter filter in one way or another. For epidemics in wildlife or domestic animals and plants, killing susceptible or infected individuals (*culling*) removes these individuals from the population entirely, hopefully minimizing subsequent transmission and thus reducing R_0 below 1. Culling is a commonly used strategy, albeit a controversial one, for controlling the foot and mouth disease virus in cattle. Post-exposure treatment (antibiotics, antivirals, etc.) increases the rate at which individuals move into the recovered compartment, shortening their infectious period and reducing the number of susceptibles they can infect; when treatments are not available, contact tracing and isolation can similarly curtail the infectious period. Finally, transmission controls such as quarantines block infection without moving individuals between compartments.

The SIR model provides a quantitative framework for calculating how much control is necessary to eradicate a disease, or how much a given level of control will reduce the level of disease in the population. Suppose we can eliminate some fraction of effective contacts, by a *control fraction* (p), by closing either the compatibility filter (e.g. by vaccination) or the encounter filter (e.g. by providing condoms or clean needles, or by social distancing). Then the value of R_0 will be reduced by a factor $1 - p$; if R_0 is initially equal to 4 and we can achieve a control fraction of 0.75 or 75 per cent, then we will reduce R_0 to $(1 - 0.75) \times 4 = 1$.

A little bit of algebra shows that in order to reduce R_0 to less than one we need to increase the control above a critical value of $p_{crit} = 1 - 1/R_0$ (Figure 2). This immediately gives us one explanation for why it was much easier to eliminate smallpox ($R_0 \approx 6$, $p_{crit} \approx 0.8$) than it has been to eliminate measles ($R_0 \approx 15$, $p_{crit} \approx 0.95$), despite

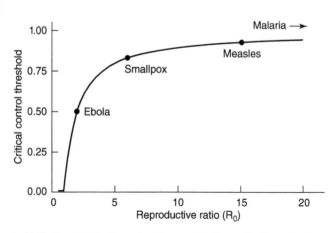

2. Critical control level required (proportion immunized or treated to prevent transmission) to eradicate an infectious disease, based on its R_0 value.

the fact that cheap and effective vaccines are available for both diseases, and why it will be extremely difficult to eradicate malaria, even once we have an effective vaccine: some researchers have estimated R_0 to be greater than 100 in some areas, so the critical control fraction would be greater than 99 per cent. In fact, the only way to eradicate malaria in high disease areas will be to combine several different strategies (such as drug treatment and mosquito control), each of which could have (say) 90 per cent effectiveness, so that their combined efficacy could reach the 99 per cent level that might be required.

If disease control measures can reduce R_0 below 1, they will not only terminate any existing epidemic, but will prevent recurrence of the epidemic as long as the control measures are maintained. Eradicating a disease within a given region, such as the UK or Europe, reduces the local burden of infectious disease, but does not eliminate the need for disease control unless public health authorities can somehow be completely sure that they can prevent the importation of disease from outside the eradication zone. Only if we can eradicate a disease globally, as has so far been done only for smallpox and rinderpest (a lethal cattle disease closely related to measles), can control measures safely be discontinued. This makes eradicating a disease, rather than simply controlling it, an attractive policy option—once the disease is completely gone, any resources that went into managing it can be freed for other disease control efforts, or for other societal goals.

Knowing R_0 does not tell us everything about controlling disease—diseases such as influenza ($R_0 \approx 2 - 3$) and HIV ($R_0 \approx 2 - 5$) are harder to control than their relatively low R_0 values would suggest. Sometimes treatments are unavailable, or too expensive. In other cases, treatment or control measures are only partly effective. With a vaccine that is only 50 per cent effective, comparable to the experimental malaria vaccines currently being tested, and better than the best HIV available (≈ 30 per cent effective against infection), twice as many people need to be treated (if $R_0 > 2$ it would be impossible to eradicate the disease with this vaccine). Another problem is that infections may be hard to detect, and thus be out of reach of disease control efforts, for either biological or cultural reasons. Biologically, some individuals (carriers) can be infected and spread a disease while showing no symptoms (asymptomatic); culturally, many diseases carry a stigma that makes people hide the fact that they are infected. During Ebola epidemics, one of the major concerns about imposing harsh control measures is that they may simply encourage people exposed to Ebola to hide from authorities. Finally, the mere fact that a disease spreads quickly—has a short *generation time*, the average time between someone getting infected and the time when they transmit the infection to others—makes it harder to control an ongoing epidemic, for two reasons. First, the epidemic spreads too rapidly in the population for epidemiologists to decide on and implement control measures. Second, if a disease transmits quickly from person to person (even if the infectious period is short, so that R_0 is not too large), epidemiologists doing contact tracing will not be able to find and isolate infected people before they have already passed on the disease to others.

Compartmental models tell us much more than the level of control necessary to eradicate disease locally or globally. They also give a simple formula for the number of people who will be affected by a disease outbreak in the absence of control, or the size of the susceptible population at equilibrium for a disease that becomes established in the population (i.e. endemic). Compartmental models have also helped epidemiologists to think about the dynamics of disease—the ways that the infected population changes over time.

For example, one of the first applications of compartmental models explained that observed multi-year cycles of measles epidemics did not necessarily mean that a new genetic type was invading every few years; rather, disease spread so fast that the susceptible population was exhausted and required several years to build up to the point where it could support another major outbreak. Similarly, mathematicians have pointed out that vaccination campaigns that fail to eradicate a disease allow the number of susceptibles in the population to build up. Even if vaccination coverage stays high, these build-ups may lead to large outbreaks several years after the beginning of the campaign. Without this dynamical insight, the outbreak could easily be interpreted as a sudden change in the effectiveness of the vaccine or the transmissibility of the disease, rather than as a straightforward consequence of a sub-critical level of control.

Within-host disease dynamics

One of the many biological details that compartmental models omit in their quest for simplicity is any description of the way that disease plays out within an individual host. In compartmental models, hosts are either infected or not; we don't keep track of the level of infection within an individual (e.g. the number of virus-infected cells or the density of the virus in the bloodstream), nor of the response of the individual's immune system to the disease.

Standard compartmental models are best for understanding small pathogens (*microparasites*) such as viruses, bacteria, fungi, and protists; because populations of these pathogens build up quickly within a host, and trigger similar immune responses in most hosts, lumping hosts into just three categories—susceptible, infected, and recovered—is a reasonable simplification. In populations infected with *macroparasites*—larger parasites such as tapeworms or ticks—the number of parasites per host (*parasite load*) varies greatly among individuals.

Mathematicians have designed more complex models that can keep track of parasite load distributions, but the micro/macroparasite distinction has also begun to blur as researchers build more elaborate microparasite models that track changes in the numbers of infected particles or cells and the level of activation of the immune system within an individual. For example, a large fraction of HIV transmission occurs within the first month of infection. If we want to understand and predict HIV epidemics, we obviously need to use models that distinguish between recently and not-so-recently infected people; we might even want to track the precise level of virus in the blood and other bodily fluids of an infected person.

Models that track both changes in the number of infected people and changes in the number of infected cells within individuals are mathematically complex—one can imagine the difficulty of keeping track of all of the virus particles within every individual in a population! Somewhat more manageable are *within-host models*, which focus on the progress of disease within a typical person, ignoring how the disease spreads among individuals. Where epidemiological models represent the progress of disease in a population, and give insight into the impact and control of disease at the population level, within-host models help us understand the dynamics of disease within a single individual.

Despite this difference in scope, epidemiological models and within-host models have striking similarities (Figure 3). We can easily adapt compartmental models for within-host models, especially for parasites such as viruses that must invade host cells in order to reproduce. Instead of assuming that infection builds up quickly and characteristically within individual hosts so that we can treat them as either susceptible or infected, we now assume that the level of infection (e.g. the number of virus particles) builds up quickly and characteristically within host *cells*. The concepts of encounter and compatibility filters are just as useful on the within-host as the within-population levels, describing how infection gets from one cell to another and what prevents or allows infection of a cell by a disease particle.

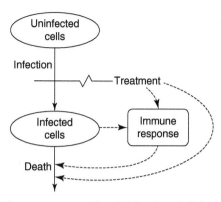

3. A within-host compartmental model showing: the infection of cells and death of infected cells; triggering of the immune response by infected cells and by treatment; killing of infected cells by the immune response and by treatment; and blocking of between-cell transmission by treatment.

Within-host models often add a compartment to keep track of free-floating infectious particles outside cells, as well as a separate variable that tracks the host's level of immune activation—for example, the number of active white blood cells of a particular type. Within-host models usually assume that immune activation increases as the number of infected cells increases. If the immune response is rapid enough and strong enough, these models show how the immune system can naturally overwhelm an infection, although not necessarily before the infection has had time to produce some infectious particles that leave the host's body and infect another host. Within-host models can also show how drug treatments can slow down disease spread within a host sufficiently for the immune response to eradicate the disease. For viruses such as HIV and the human T-lymphotropic virus that attack immune cells, within-host models show exactly how these diseases pervert the normal immune strategy; the immune system responds to the presence of virus infection by activating more immune cells, which in turn provide more resources for virus growth. This is like trying to put out a fire with gasoline instead of water.

Virulence, resistance, and tolerance

Compartmental models have been used most often for widespread diseases where nearly everyone in the population is equally susceptible, such as measles, polio, or smallpox. Humans do vary in their susceptibility to infection: because they have different genotypes (i.e. complete sets of genetic material), or are better or worse nourished, or are more or less stressed. They also vary in infectiousness, how badly they suffer, and how likely they are to die from the disease. However, for the purposes of epidemiological planning it's often wise to ignore these details, at least initially.

When we turn to thinking about evolution, however, this variation becomes central to the questions we are asking. In the last few decades, epidemiological modellers have turned from trying to understand how diseases spread in populations over timescales from days to years, to trying to understand how diseases evolve over timescales from years to thousands of years. What is it about the combination of a particular host, a particular parasite, and a particular environment that allows the parasite to infect a host? What determines whether the host is badly harmed by the infection or only has mild symptoms?

We have to make several important distinctions about the characteristics of parasites, the first of which is between infectiousness or transmissibility (how easily the parasite can infect the host) and virulence (how severely it affects the host if it succeeds). We often treat infectiousness and virulence as fixed properties of a parasite. Smallpox has much nastier symptoms than measles, and a much higher chance of killing the host, regardless of the particular genetic make-up of the parasite or of the host it infects. Measles is always more infectious than smallpox, which is more infectious than HIV or Ebola. Virulence can depend on the host as well—someone with one copy of the sickle-cell allele will suffer less from a malaria infection—and, in general, the outcome of a host–parasite interaction depends on both participants. We can imagine two parasite strains and two kinds of hosts that 'cross over' in their effects, with one parasite having higher virulence on the first host genotype and the other having higher virulence on the second host genotype.

Hosts can control the infectivity and virulence of the pathogens attacking them in two ways. If the host can close the compatibility filter partially or completely, we say that it *resists* the parasite. The parasite might not be able to infect such a resistant host at all, or its population size within the host might be limited, so that the host suffers few ill effects. Alternatively, the host might allow the parasite to infect it (or more precisely it might not invest energy in defending itself), but it could evolve mechanisms so that it suffered less harm from infection: in this case, we would call the host *tolerant* rather than resistant.

Tolerance and resistance lead to similar outcomes at the level of the individual host (the host isn't harmed by the parasite), but very different outcomes at the level of the population. If some individuals are highly susceptible (neither resistant nor tolerant), then the presence of resistant individuals will help them by lowering the overall chances of infection, while tolerant individuals will increase the chance of infection. This is one reason that epidemiologists worry about the introduction of partially effective vaccines. If pathogens evolved to replicate more quickly within the host in order to overcome partial resistance in vaccinated people, they might increase their virulence in non-vaccinated people; if vaccination makes people tolerant rather than resistant to disease, they could still spread infection to unvaccinated people.

Chapter 3
Influenza

Unless you are very lucky or very careful, you have probably had influenza (the flu) sometime in your life. Flu is familiar to almost everyone, occurring in every country, every year. It has been with us throughout history. Though not as gruesome as Ebola, the flu virus has caused more deaths than any single disease outbreak since the Black Death (bubonic plague) of the 14th century: 20 to 50 million people worldwide died from the 1918 Spanish Flu.

Furthermore, although non-epidemiologists may not think of it as a big problem, the annual flu epidemic that occurs every winter in temperate parts of the world infects millions of people. Although influenza kills only a small fraction of even the frailest elderly population, it is still thought to cause as many as 40,000 deaths in the USA in a typical year (not a pandemic year). Because influenza causes many deaths indirectly, for example due to secondary infections, these numbers are quantified indirectly by estimating how many 'excess' (above those expected from other causes) deaths are observed.

Influenza is even scarier in years, such as 2009, when we think we might be on the verge of a deadly pandemic that could cause millions rather than tens of thousands of deaths. Pandemics arise under very particular circumstances, but the two most important ingredients are lack of existing immunity (an opening in the compatibility filter) and virulence. If the virus has radically changed its appearance, people will be more susceptible than average because their immunity from previous years fails to protect them against the new strain, making the value of R_0 (the intrinsic reproductive number, see Chapter 2) higher than in a typical year. Higher transmission is even scarier if the new strain is also unusually virulent, severely harming a larger fraction of its victims. Authorities feared this situation in 2009 for three reasons: (1) the new strain was of the H1N1 type, different from the previous year's H3N2 type, increasing susceptibility and driving up R_0; (2) an H1N1 strain caused the highly virulent 1918 pandemic; (3) the new strain's virulence was initially overestimated as the average severity of cases in Mexico, the new strain's origin, was overreported at first.

In the end the 2009 strain's virulence turned out to be about average, although it did affect younger people relatively severely, leading to more years of life lost. The 2009 H1N1 outbreak was officially classified as a pandemic—that is, it was caused by a previously unobserved strain that caused significant numbers of cases all over the world—but happily it infected fewer people, and killed fewer of them, than initially feared.

To prevent pandemics, we have to control transmission. Transmission can be controlled by reducing encounters (e.g. by sneezing into your elbow instead of into your hand), by reducing compatibility (vaccinating to reduce the number of susceptible individuals), or ideally by a combination of both. In the 2009 H1N1 epidemic, encounter rates were reduced by closing schools throughout Mexico, as well as by discouraging large public gatherings and distributing masks and hand sanitizer. A vaccine was swiftly developed to close the compatibility filter for the new strain; the vaccine became available in October 2009, a mere six months after the strain was characterized. Vaccines were initially limited, so were first distributed to the target groups believed to be most at risk, *and* to those most likely to transmit the disease, including school-age children. We'll come back to why children are an important target group for flu vaccination later in this chapter.

In high income countries, the flu shot (or at least the ubiquitous publicity surrounding the vaccine) is as much a harbinger of winter as the shortening of the days. Unlike the measles or diphtheria vaccines that children get once or twice in high income countries, we need new flu shots every year because the flu virus evolves rapidly; it changes its outer garments so fast that our immune system needs new clues each year to recognize the current disguise of this old and otherwise familiar foe.

To understand flu control, it helps to understand the evolutionary processes that lead to flu's unique capacity for costume changes. As discussed in Chapter 1, hosts and parasites are like Alice and the Red Queen in *Through the Looking Glass*: they have to run as fast as they can to stay in the same place. Hosts and parasites are locked in a race, with the host evolving to escape the parasite and the parasite counter-evolving to keep up with the host. For the host, winning the race means closing the compatibility filter so that the parasite can no longer exploit it. For the parasite, winning means keeping the compatibility filter open, so it can continue to exploit the host. Thus, 'the same place' means the parasite can still infect the host; running as fast as they can means both the host and parasite are rapidly evolving, with the former trying to set itself free and the latter stubbornly holding on.

Evolution does not always mean evolution via natural selection, the 'survival of the fittest' paradigm that you may remember from school. Any natural population contains many different genotypes; the proportion of any given genotype within the population is called the *genotype frequency*. To evolutionary biologists, evolution means *any* change in genotype frequencies over time. Popular discussions of evolution mostly focus on the process of natural selection, the change in the frequencies of genotypes because of differences in fitness—that is, the expected numbers of offspring of each genotype. But genotype frequencies can also change due to chance events, a process referred to as *genetic drift*.

To distinguish further between natural selection and genetic drift, let's consider an imaginary infectious disease. Suppose a host genotype randomly mutates to become completely resistant to the disease, with no bad side effects. The mutation should increase in frequency in the population due to natural selection; people who carry the mutation will have higher fitness in the presence of the disease, and the same fitness in its absence. However, there's a small probability that the mutation will be lost before it has a chance to increase in frequency: for example, if the extended family in which our resistance mutation arose decides to embark on a trip together—and is then wiped out in a bus crash. This tragic scenario is an example of genetic drift: evolution happening because of a chance event that has nothing to do with the infectious disease, the host's fitness, or the mutation itself. Predicting flu epidemics is difficult even for expert epidemiologists in part because flu evolves by genetic drift as well as by natural selection. Chance events that happen before the flu season can set the year's flu epidemic off down different, unpredictable tracks.

Evolution requires genetic variation. The ultimate source of genetic variation is mutation, a process at which viruses excel. Genetic variation also arises through recombination, when existing genotypes get mixed up in different ways. The influenza virus takes advantage of both mutation and *reassortment*, a type of recombination.

Relative to influenza (and most other pathogens), humans have lower mutation rates, smaller population sizes, and longer generation times. These characteristics limit humans' ability to evolve, but we make good use of recombination as a key feature of our immune system. The reason we can recognize so many

31

parasites is not because our genes code a different protein (*antibody*) to recognize each one—our genomes would have to be many times their already huge size. Instead, we reuse the same small pieces of genes in different combinations to create a range of antibodies. These different antibodies recognize specific parts of the costumes, which are known as *antigens*, of many different parasites.

Our genomes are made of DNA that is replicated or copied by a protein called a *polymerase*. Our DNA polymerase not only copies DNA, it also proofreads the new strand being produced, and can correct many of the inevitable errors made during the copying process. The influenza virus has a genome coded in RNA rather than DNA. It also encodes its own RNA polymerase. However, unlike the polymerase that we produce to replicate our DNA, flu's RNA polymerase cannot proofread to correct replication errors. As a result, replicating flu viruses end up with many more errors—that is, mutations—in the new copies of the genome. The virus arising from the new, mutated RNA genome will inherit the mutation, and so will its offspring. As a result, flu's mutation rate is 100,000 times greater than our own. Because flu mutates rapidly, it evolves rapidly.

The accumulation of mutations in influenza leads to *antigenic drift*, a process of slow change in the flu virus over time (note, antigenic drift is different from the genetic drift process discussed earlier). Under antigenic drift, our bodies can sometimes use the same set of antibodies to recognize strains of flu that have drifted apart. Their short-sleeved shirts may have changed from stripes to solids to plaids, but they are still recognizable as shirts. The new varieties of influenza that result from antigenic drift then primarily evolve through selection—genotypes that are slightly better than average at evading the existing repertoire of host antibodies will have higher fitness. Because flu constantly evolves, both vaccinated people and unvaccinated people who contract flu tend to become more susceptible after a few years.

The flu has another evolutionary trick. Its genome is divided into eight discrete segments; it can combine different variants of the segments in new ways to result in viruses with novel properties. When two different variants of the flu virus happen to invade the same cell within a host, they can reassort, trading segments of their genomes with one another to produce an *antigenic shift*. This reassortment results in fast, dramatic costume changes, effectively trading a short-sleeved shirt for a smoking jacket. These changes are hard for our immune systems to recognize, so they open the compatibility filter. As a result, the reassorted offspring have an evolutionary advantage and spread in the population—but only if they are also good at transmitting themselves to other hosts, overcoming the encounter filter.

Evolution caused by antigenic drift does not cause pandemics. The hallmark of pandemic flu is antigenic shift, particularly the reassortment of flu strains found in multiple hosts. For example, the 2009 H1N1 flu arose in pigs (hence the term 'swine flu') when different segments from viruses that were adapted to birds, humans, and pigs came together.

The constantly changing costume of the flu virus consists of a protein derived from two separate genes located on two segments of the genome: haemagglutinin (HA) and neuraminidase (NA). Flu strains are named for their variants of each gene: H1N1 combines HA variant #1 with NA variant #1. These swaps contribute to antigenic shifts, but shifts also happen when reassortment happens within a subtype. When a given strain of flu swaps its HA and NA genes for another type, our immune system is less able, and sometimes completely unable, to recognize the flu virus. HA and NA are antigens, the parts of a parasite that are recognized by our immune system. Antigens are exposed on the surface of virus particles, where our immune system can detect them. When viruses infect a cell, they turn the cell into a factory for the production of more viruses; in the case of flu, the new viruses emerge from the cell by budding out, rather than bursting the cell altogether. As part of the budding process, infected cells display HA and NA on their surfaces, triggering the recognition and destruction of the co-opted cells by our immune system.

There are two major types of flu vaccine. The standard flu shot is an intramuscular injection of three of four inactivated or 'killed' virus strains. A less widely used type of vaccine is a nasal spray of 'live' attenuated (i.e. weakened) influenza virus (LAIV). (We put 'killed' and 'live' in quotes here because the consensus among biologists is that viruses are not living organisms, although the terms 'live' and 'killed' are still commonly used in the vaccine literature.) Both types contain a mixture of three different strains of virus that are predicted to be common during the upcoming flu season. Sometimes some of the same strains are included in consecutive years, if they are still common, but vaccine developers usually include at least one new strain as well, in the hopes of anticipating antigenic drift or shift.

Both vaccines trigger a response in our most common type of antibody, immunoglobulin G (IgG). A given form of IgG can handle the relatively subtle changes that can occur as a result of antigenic drift within a year, but not larger changes. In other words, if the injection contained only versions of the virus with stripes, and mutations occur that change the costume to a solid colour, our immune system won't be able to detect the changed virus—let alone more complex changes resulting from antigenic shift, that is, changing T-shirts to smoking jackets. It is this relatively limited ability of IgG to deal with differences in HA and NA, combined with the continuing process of antigenic drift, that requires us to get a new injection with a newly developed vaccine each year.

If it is so effective, why isn't LAIV given more widely? There are three major reasons. First, LAIV requires cold storage, making supply chains much more difficult. Second, the injection is more effective in generating an immune response in adults.

Third, people who receive LAIV can *shed* virus (i.e. release it into the environment where it can infect others) for a short period after vaccination. This can be beneficial as unvaccinated people who come into contact with someone who is shedding become immune as well, but it also could be risky for people with weak immune systems who come in close contact with recently vaccinated people. LAIV is not recommended for people with weakened immune systems, including pregnant women, people under 2 or over 49, and people with chronic infections such as HIV or tuberculosis.

Modern vaccination programmes try to go beyond protecting individuals to protect the entire population. Vaccines can reduce the severity of the annual flu epidemic in several ways. First, they can prevent infection of the vaccinated person altogether. This clearly means less infection—we protect not only our vaccinated individual, but also anyone she would have transmitted disease to if she had been infected. However, even vaccines that do not completely block infection can reduce subsequent transmission. Adults who have received injected vaccine usually recover faster than unvaccinated people even when they are unlucky enough to get the flu. This is good for individuals (they suffer symptoms for a shorter time and may be able to get back to work sooner); it also cuts down on overall infection because they have less chance to transmit to others during their shortened infectious period. Vaccinated people may also shed less virus, and hence be less infectious, during the time the virus is present in their bodies. Finally, vaccinated people help control the epidemic via herd immunity (see Chapter 2): potentially infectious contacts of infected people with vaccinated people are wasted (from the pathogen's point of view), leading to a reduction in R_0. If we can successfully immunize enough people (specifically, more than a fraction $[1 - 1 / R_0]$ of the population), then infectious people will generate fewer than one new case each and the virus will go extinct.

Vaccination prioritization strategies are important, particularly if vaccine supplies are limited, as was the case early in the 2009

H1N1 outbreak. Even when there is enough vaccine to go around, it is essential for public health agencies to target their advertising and outreach, given that resources are finite and that people are inundated with information. One obvious choice is to concentrate on people who are most at risk of severe disease or death. However, many of these people are vulnerable precisely because they have weakened immune systems, which means that vaccination may not protect them even if they can be convinced to get the shot. Another approach is to vaccinate those most likely to be exposed: for example, healthcare workers. A third, complementary approach is to focus on vaccinating the people who are most likely to transmit the virus (this approach might also prioritize healthcare workers, who are frequently in contact with vulnerable people).

Epidemiologists use contact networks to understand how disease spreads within a population. Contact networks focus on the 'encounter' stage of transmission—if an uninfected person doesn't encounter an infected one, transmission won't happen. Figure 4 shows a middle-class family in the USA, consisting of two parents, two children, and one grandparent. One parent works outside the home in a small business with three other co-workers, while the other parent has an online business and spends a lot of time working in the local coffee shop with a business partner. The two children, aged 5 and 7, attend the local elementary school. Each child has 10 classmates plus one teacher. The grandparent lives independently but near the family's home, and comes over for dinner several times a week.

Casual interactions like saying hello to people only occasionally result in flu transmission. The most important encounters for flu involve regular physical contact, or opportunities for sneezing on and being sneezed upon. Each such contact, for each member of our core family of four, is shown as a line in the figure. This simplified (but not unrealistic) example shows that the children have the highest number of epidemiologically meaningful

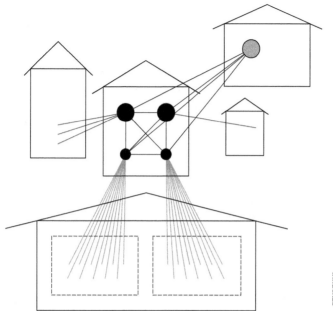

4. Contact network for a Western middle-class family consisting of two parents, two school-age children, and one grandparent living independently.

encounters. By extension, they are the most likely to bring disease into the family, particularly to the grandparent, who is the most likely to have complications from the flu. For this reason, vaccinating kids is a high priority not only to protect them, but also to protect the elderly. Contact networks vary considerably from one region to another, as well as by socioeconomic status within regions, so careful comparative work is required to set global vaccination priorities.

Flu also represents an interesting case study for an age-old debate: are some kinds of research too dangerous to allow? If people do go ahead and conduct this research, should the results be available only to a limited, carefully screened subset of researchers and policymakers, or should scientific knowledge be available to all?

Scientific research is funded by governmental organizations, non-governmental organizations (NGOs), and industry. Industrial research is usually proprietary—it is kept private so that companies can get a return on their investment. But research funded by governments and NGOs is science for the people. The results of such work should be available to all, particularly when the research is ultimately paid for by taxes and charitable contributions and is directly relevant to human health.

Unfortunately, some of the same research that helps predict pandemics could also be used for nefarious purposes by bioterrorists. Such 'dual-use' research is subject to international treaties and export restrictions. Even if it is not deliberately misused, terrible things could happen if laboratory-developed super-strains accidentally escape the confines of research labs. The most critical indicator of pandemic flu—what determines whether a new combination of HA and NA types from avian, porcine, or human sources can spread widely in human populations—is the route of transmission. Flu becomes pandemic only if efficient human-to-human transmission is achieved—specifically, airborne transmission. Thus, an important public health priority might be to understand how airborne transmission between humans could evolve and what hallmarks we should be looking for in order to be able to detect emerging pandemic strains before they spread very far.

Flu research during the first two decades of the 21st century resulted in a face-off between people who embraced the democratization of knowledge and people who feared misuse of that knowledge. The primary motivation behind the controversial research is simple: if we can understand what genetic changes make pandemic flu different from routine seasonal flu, specifically what changes are required to allow airborne transmission, we could potentially see a pandemic strain coming and take appropriate precautions. Maximum response to every case of flu is not an option, because resources are limited and because we value individual freedoms as well as public health. Thus, extraordinary control measures should only be taken when risk is high, but how can we know when that threshold has been crossed?

Flu research became extremely visible in 2011 not because of breakthroughs in understanding pandemics, but because of controversy over the publication of two high profile papers about airborne transmission. Scientists were studying how H5N1, a particularly virulent strain of avian flu, could achieve airborne transmission from one ferret to another. Ferrets are an important model animal system for studying flu transmission in humans, in part because, like people, ferrets sneeze when they are infected with flu.

The road to publication of this research was long and rocky. The papers were originally submitted in August of 2011. Four months later, the US National Security Advisory Board for Biosecurity initially recommended that key details of the experiments be omitted from the two papers to address security concerns. In May of 2012, the first paper was finally published online, having languished for two months following acceptance. The second paper came out in late June of the same year. The path to uncensored publication was cleared once experts around the world decided that the potential public health benefits of publication, with all the details, outweighed the potential harm.

Another potential risk of this type of *gain of function* research—research that tries to create more virulent or more transmissible forms of human pathogens—is the possible accidental release, or escape, of the lab-evolved strains into the community. In 2012, scientists and politicians were more worried about terrorism than about the possibility of escape; the debate over the origins of the COVID-19 pandemic (Chapter 8) has shown how important lab escapes are from both political and public health perspectives.

One particularly interesting aspect of the H5N1 controversy is that in the winter of 2012 scientists themselves agreed to a moratorium on gain of function research until concerns and proper safeguards could be discussed. The 2012 moratorium had initially been proposed to last for 60 days, but ultimately lasted almost a year. Such self-imposed restrictions are rare, and are critical to maintaining public confidence in science. However, in October 2014 the US government issued a new moratorium banning funding of new research on gain of function, not because of fears of bioterrorism but because it was discovered that the key US government research centres had mishandled potentially dangerous samples of other pathogens: fears of escape were clearly not unfounded. The ban on funding was lifted in 2017, following implementation of new safety protocols, commitments to transparency, and special considerations for research on potential pandemic pathogens.

The 'two most famous papers almost not published' found that HA, one of the two most important garments involved in flu's costume changes, determines both how transmissible *and* how virulent a particular strain of flu is likely to be. For example, some forms of HA such as those found in H1N1 can infect human cells via proteins found on the surface of cells in our noses and throats. These forms are highly transmissible, because the virus can easily find its way in and out of new hosts when it doesn't have to travel very far into our bodies. But H1N1 is not highly virulent, because the virus does not usually find suitable cells to infect deep in our lungs, and is thus less often associated with pneumonia. Other forms of HA, such as H5N1, encounter appropriate cells only deep in the lungs. They can therefore cause damage leading to pneumonia, and so are far more virulent. However, transmission of this dangerous strain is low because H5N1 particles can't infect us unless they find their way far down into our lungs. This tradeoff is also observed with SARS-CoV-2, the causative agent in the current COVID-19 pandemic.

Thanks to the now published research on ferrets, we know that H5N1 only needs a small number of mutations to evolve the capability for airborne transmission between humans—at least for the particular strain of H5N1 virus that the researchers studied. Moreover, we are pretty sure what those mutations are. We have identified the costume change that this particular strain of virus would likely perform if it were to become pandemic—and thus we could recognize it *before* a major outbreak. If these signature mutations are found in all airborne strains, then we can detect when flu is on the path to becoming airborne (and hence likely pandemic) by tracking the genetic sequences of flu samples from domestic poultry and human patients.

Why the caveats about 'this particular virus strain'? Because influenza evolution, like any other kind of evolution, depends on where you start. While we understand how the particular strain of H5N1 that was studied by the researchers could become pandemic, we don't know if we can generalize those ideas even to other variants of H5N1, let alone to very different strains that have caused pandemics (the 1918 flu was H1N1; the 1968 or Hong Kong flu was H3N2; etc.).

Regardless of one's personal opinion about the wisdom or the utility of the 'two most famous papers almost not published', they have set important precedents for future dual-use research. The restraint that scientists used while trying to decide whether to publish this work suggests that the research community, as well as world leaders, may at last be embracing the precautionary principle (scrutiny prior to any negative consequences, rather than after the fact). It also suggests that scientists realize that the active engagement of a broad array of stakeholders is essential to maintaining public trust in science.

Chapter 4
HIV/AIDS

Our second case study is HIV, another virus likely to be familiar to most readers. HIV is the human immunodeficiency virus that causes acquired immunodeficiency syndrome, or AIDS. Once a person is infected with HIV, they may live for many years without showing any symptoms, but they can still transmit the virus. If untreated, the virus population within their body will explode after several years, ultimately causing the person's immune system to collapse. They then become vulnerable to opportunistic infections (see Chapter 2) by disease agents such as the fungi *Pneumocystis jirovecii* or *Candida albicans*, neither of which can infect people with normally functioning immune systems. Certain types of cancer, such as Kaposi's sarcoma, also occur more frequently in people living with AIDS. Untreated, HIV infections are usually fatal within 5 to 10 years. Note that people with AIDS die from opportunistic infections, rather than from HIV itself.

HIV is very different from influenza. HIV is transmitted in our most intimate moments, by exchange of bodily fluids. Many people think of HIV solely as a sexually transmitted disease, because both semen and vaginal fluids carry enough virus to cause a new infection. However, because our blood also carries HIV, transmission can happen any time someone comes in contact with an infected person's blood. Medical personnel stick themselves with needles accidentally, and intravenous (IV) drug users share needles for economic or social reasons. Recipients of blood transfusions and blood products, especially haemophiliacs and others who need frequent transfusions, were often infected before routine, rigorous screening for HIV and many other blood-borne pathogens was instituted. Because breast milk also carries the virus, babies can contract HIV while nursing from their HIV-infected mothers.

Understanding routes of transmission is key to protecting both populations and individuals, who can modify their behaviour accordingly (safe sex, needle exchange programmes for IV drug users, eye protection and needle stick protocols for health workers, etc.). As we'll see later, identifying the most common routes of transmission also helps inform the use of limited resources for HIV treatment. But before we can understand treatment, we need to know more about HIV.

One of the biggest challenges of HIV is its extraordinary evolutionary potential. One might almost believe that HIV has a cloak of invisibility, rather than a series of costumes like influenza. The first few decades of research on HIV were nightmarish, because of HIV's uncanny ability to become invisible to our immune systems. We could not develop vaccines, because the virus changed too quickly—much more quickly than influenza, whose costume changes on the scale of years are problematic enough! That evolutionary prowess also means that HIV is enormously variable. So, in trying to create a vaccine, we would not be tracking just one moving target, but many. We feared we would never be able to develop a drug with lasting efficacy, because HIV seemed to effortlessly become invisible to our medicines as well. In order to understand HIV, then, we have to understand the details of how its pernicious disappearing act works.

Part of HIV's invisibility comes through an insidious strategy: it hides inside our own genomes. HIV is a retrovirus, which means that its genome is coded in RNA that is copied into DNA by a viral enzyme (i.e. a chemically active protein), *reverse transcriptase* (RT). The resulting double-stranded DNA can then be integrated into our genomes by means of another viral enzyme, *integrase*. The fact that HIV can become 'us' is one of the reasons it is so difficult to cure, as inactive copies of HIV can lurk within our genome and later become reactivated.

However, HIV can only enter certain types of cells, an example of the compatibility filter described in Chapter 2. In order for HIV to enter a cell, the cell must have surface proteins (receptors) that fit a particular knob sticking out of HIV's envelope, *gp120* (glycoprotein—a protein with sugars attached to it—with a molecular weight of 120). Thus, blocking the gp120 compatibility filter should confer resistance to HIV. Remarkably, there is a mutation in humans that does almost exactly that. HIV actually uses two receptors to enter the cell: the primary receptor for gp120, which is called CD4, and a co-receptor, called CCR5. Humans with the CCR5-Δ32 mutation lack the co-receptor for gp120 and are hence resistant to HIV infection. However, for complete protection from HIV, a person must have two copies of this mutation, one on each strand of their DNA. People with a single copy of CCR5-Δ32 (heterozygotes) can still be infected and become symptomatic, though they are more resistant than people with no copies at all.

While we might guess that the CCR5-Δ32 evolved to protect humans from HIV, it was already present in human populations long before the HIV epidemic. HIV has likely been infecting humans only for about a century, but we have detected the CCR5-Δ32 mutation in Bronze Age skeletons from 3,000 years ago. Because HIV is much younger than that, and because the CCR5-Δ32 mutation is common in some human populations, researchers speculated that CCR5-Δ32 might have evolved to protect against some older pathogen, such as bubonic plague or smallpox. However, careful analyses of the CCR5-Δ32 gene and its surrounding DNA revealed that the high frequencies today are most likely a happy accident—the result of genetic drift rather than natural selection. Other, less well-known genes conferring resistance to HIV also appear to be far more ancient than HIV.

Only three human cell types have receptors and co-receptors that bind with gp120 (and hence can be invaded by HIV); all three are part of our immune system. It is because HIV targets immune cells that infection by HIV results in immunodeficiency. The virus hides in these cells until it is activated. Once activated, the virus begins to replicate, ultimately resulting in the death of these immune cells and weakening our ability to respond to other infections.

The mutation rate of a virus is critical to understanding how easily (and hence how frequently) any given mutation can occur (as noted in Chapter 3 on influenza, mutations happen at random, without respect to whether or not they are going to be advantageous to the virus). Intuitively, we might expect HIV to have a much higher mutation rate than influenza since it evolves so quickly. And HIV's mutation rate is indeed more than 10 times higher than influenza's.

HIV replication uses two distinct enzymes. One is the virus's own RT, which as described previously converts viral RNA into DNA. The DNA versions of the viral genome which have been integrated into our own DNA are copied by a second enzyme, our own RNA polymerase. These new RNA genomes are then repackaged into the virus particles that emerge from one cell to infect another. Most HIV mutations come from the reverse transcription step. The HIV RT enzyme makes more mistakes than the enzyme used by influenza, so HIV's mutation rate is higher than that of flu.

Mutation is not the only source of variation for HIV. Unlike influenza, which has a segmented genome and so can combine different viral segments from multiple viruses, HIV has a non-segmented genome, but each *virion* or virus particle contains two copies of the genome. When two HIV viruses with different genomes infect the same cell, one copy of each can end up in the same virion, allowing their genetic material to get mixed up through a process called recombination. HIV's recombination happens during the reverse transcription step, when the information in the RNA is copied into DNA. The viral RT sometimes switches from one template, the viral genome being copied, to that of the second viral genome in the virion (and back again). This process is called template switching. To understand template switching, imagine that you were using tracing paper to trace two parallel lines (representing the two HIV genomes within one virion), but you are only allowed to trace one line at a time. You start at the left end of one line and move to the right. Now imagine that your tracing paper shifts up or down at random, such that sometimes you are tracing the top line and sometimes you are tracing the bottom line, but always drawing from left to right. In the end, you will have drawn a single straight line, but it will contain copies of parts of both lines. Similarly, recombinant HIV genomes contain all the parts of a single genome, but created from two different versions.

Mutation and recombination constantly change HIV's genome, to the point that it becomes nearly invisible to our immune system. Because vaccines rely on an immune 'photographic memory' of a pathogen target, designing a vaccine against HIV has proved extraordinarily difficult; the best HIV vaccines tested so far only prevented disease 60 per cent of the time after 1 year, decreasing to 31 per cent after 3.5 years. But HIV's invisibility is perhaps most notorious in its astonishing ability to render drugs useless, because it evolves resistance so quickly within an individual host.

In contrast to mutations in the human genome like CCR5-Δ32 that enable us to resist HIV, HIV's resistance to drug therapies usually arises from new virus mutations within individual human hosts, though resistant strains are sometimes transmitted from person to person. Despite these challenges, a regimen of *highly active anti-retroviral therapy* (HAART), developed in the mid-1990s, is extraordinarily effective against HIV. HAART uses a combination, or 'cocktail', of drugs to reduce HIV proliferation. Some of the drugs work to directly block HIV reverse transcription, in two different ways. One class of drugs acts by tricking the RT enzyme into incorporating chemicals that stop the extension of the RNA genome. These chemicals look like the building blocks used by RT, but function differently. Once a replicating HIV genome incorporates one of these chemicals, the enzyme can't continue synthesizing the genome. These drugs are called *nucleoside reverse transcriptase inhibitors* (NRTIs). Another class of drugs directly binds to the RT, stopping it from working. These drugs are called *non-nucleoside reverse transcriptase inhibitors* (NNRTI); usually only one NNRTI is used in HAART to complement the NRTI. Still other HAART components inhibit other stages of the viral life cycle.

The mutations that make HIV resistant to NNRTI (for example) are extremely improbable. However, even an event that is unlikely on a case-by-case basis can become common if it has sufficient opportunities to happen. Individual HIV genomes don't last long inside a human: a given HIV virion in the bloodstream survives for about six hours. Because the total number of virions is approximately constant, that must mean that virions are replaced by new ones about four times per day. Each person infected with HIV has tens of millions of virions in their body; thus there are hundreds of millions of cycles of replication, and hence opportunities for mutation, every few days. We can use a lottery analogy: if you buy hundreds of millions of lottery tickets, you have great chances of winning the jackpot. Because there are so many viruses turning over so quickly within an infected individual, even the most unlikely mutations will happen eventually.

The success of HAART rests on two principles: first, that mutations happen at random; and second, the basic rule of probability about the co-occurrence of independent events. In order for the virus to evade HAART, it needs three distinct mutations, one for each of the three drugs in the cocktail. Because the mutations occur independently, the chance of their happening in the same individual is the product of their independent probabilities. That is, if the chance of each event were 0.5, then the chance of two such events would be 0.5×0.5, and for three, $0.5 \times 0.5 \times 0.5$. In reality, the probability of each individual mutation happening is much, much smaller than 0.5. Instead of 'extremely improbable', we are dealing with 'extremely improbable cubed', so that even with tens of millions of viruses and hundreds of millions of replication cycles, resistance does not arise for a very long time, if ever.

Anti-retroviral therapy was once reserved for patients with full-blown AIDS, in part because of the problems of drug resistance and limited resources. However, in 2013 an exciting new paradigm was announced by the World Health Organization: 'treatment as prevention'. Under this strategy, infected but otherwise healthy patients are given treatment to help reduce transmission. People on effective drug therapy have extremely low levels of virus in their bodily fluids, so are very unlikely to transmit the virus. They also remain healthy, happy, and productive for much longer; many never even become symptomatic.

Taking the idea of 'treatment as prevention' one step further, we also have pharmaceutical tools to prevent infection in the first place, called *pre-exposure prophylaxis* (PrEP). PrEP can either take the form of a daily tablet containing a drug that blocks reverse transcriptase or a bi-monthly injection that blocks the active site of the viral integrase. PrEP is only recommended for people at high risk of infection, such as those whose sexual partners are HIV positive.

Like most medications, PrEP and HAART only work if people take their pills every day. Otherwise PrEP recipients will again be at increased risk, and in people with HIV infections who discontinue HAART the virus will start to increase again, because of the copies of the virus hiding inside our own cells that are not affected by HAART. Indeed, interrupting therapy quickly results in virus numbers rebounding to pre-treatment levels. In order to completely cure an HIV infection, then, we would have to eradicate all the cells in our own bodies whose genomes unwittingly harbour HIV.

Researchers have considered various strategies to expunge the latent copies of HIV from infected people's bodies in order to cure them. The most radical solution, and the only one that has ever succeeded, involves using radiation or chemotherapy to destroy the bone marrow cells (where the three cell types containing latent HIV reside) of a patient who is suffering from leukemia as well as being infected with HIV. Doctors then use a transfusion to replace the destroyed cells with healthy cells from an uninfected donor, while simultaneously using aggressive antiviral therapy to destroy any viruses currently circulating in the blood. To date, this type of treatment has apparently cured three patients (known as the Berlin, Düsseldorf, and London patients) of HIV. The Berlin patient survived without detectable HIV for 13 years, until death from cancer recurrence. As of 2023, Adam Castillejo (formerly known as the London patient) has been HIV-free (and cancer-free) for over 5 years; the Düsseldorf patient has been HIV-free for over 9 years. However, as well as being extremely expensive, these treatments have only ever been tried in HIV-positive cancer patients. Nevertheless, these successes continue to inspire the search for practical cures that either eradicate HIV from the infected person's body completely or allow their immune systems to suppress HIV concentrations to low levels without the use of drugs. (Another reason to try to cure HIV rather than relying on HAART is that about 50 per cent of people living with HIV suffer from some level of neurological impairment (HIV-Associated Neurocognitive Disorder, or HAND), even when HAART is working to lower their virus load to undetectable levels.)

More practical, short-term approaches to controlling the HIV epidemic rest on the less glamorous population-level and behavioural strategies of (1) promoting condom use and male circumcision, both of which reduce transmission; (2) pre-exposure prophylaxis (PrEP) for members of high risk populations; and (3) finding, diagnosing, and treating HIV-positive people. The UN's programme on HIV/AIDS has defined a '90/90/90' target: 90 per cent of people with HIV should know they are HIV-positive; 90 per cent of people who know their status should be receiving HAART; and 90 per cent of the people in treatment should have their viral loads reduced to a low level. While some countries have achieved this goal (which was supposed to be reached worldwide by 2020), progress is disappointing in many other countries, and has been further set back by disruption of health services due to COVID-19.

An important question in HIV research, and about emerging diseases in general, is 'why now'? Where did HIV come from, and how did it so quickly become a worldwide threat to health and economic stability? If we can understand the answers to these questions, we may be better able to prevent or at least slow down the next emerging disease.

Researchers have used two primary tools to understand the emergence of HIV: contact networks (discussed in Chapter 3) and phylogenetic trees. A phylogenetic tree is one of the fundamental tools of evolutionary biology. It is a way of grouping organisms to show their ancestral relationships—essentially a family tree of some group of organisms. Figure 5 shows the phylogenetic trees for HIV and for influenza. Nowadays, most phylogenies are built using genetic information. The more recently two organisms share a common ancestor, the more closely related they are, and the

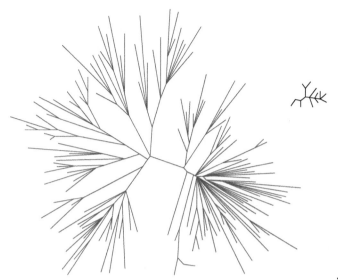

5. HIV (left, larger) is characterized by extensive genetic variation at any given time, with strains persisting for long periods of time. Influenza (right, smaller) is characterized by little variation at any given time, and strains replacing one another, rather than coexisting, over time.

more similar their genomes will be. The tips of a phylogenetic tree representing closely related genomes are close together, while more distantly related genomes are further apart. Our ability to build phylogenetic trees relies, once again, on the random nature (and relatively constant rate) of mutation: genome similarity is assessed in terms of numbers of mutations that differ between two organisms. But we can do more with phylogenetic trees than just understand relatedness. By analysing the shape of a phylogenetic tree, we can understand what kinds of evolutionary processes have occurred, and over what time scales.

If mutations happen at similar rates in close relatives, which is often true, then we can estimate time simply by counting the number of mutations. Two viruses that differ by seven mutations are more closely related, and share a common ancestor more recently, than two viruses of the same type that differ by, say, 20 mutations. The lengths of the branches in a phylogenetic tree are proportional to the number of mutations, and hence to evolutionary time.

Tree shape can also tell us a lot about virus evolution. Comparing an influenza tree with an HIV tree (Figure 5) reveals huge differences in their evolutionary patterns. The tree of influenza has a few, short branches close to the trunk. This shape reflects the fact that most strains of the flu quickly go extinct and are replaced by new varieties. HIV trees are much more complex. It's difficult to identify a core trunk because there are so many long branches, representing new strains that coexist with old strains for a long time, evolving independently.

Our best guess is that the ancestor of HIV as we know it today was contracted by some unlucky human from an infected chimpanzee butchered for food. But why did HIV become pandemic *now* (i.e. within the last 40 years)? Simian immunodeficiency viruses (SIV), the group of viruses ancestral to HIV, have been around for thousands of years, so people have probably been exposed to them for a long time. Phylogenetic methods using both contemporary viruses and old blood samples that have been carefully stored in hospital freezers tell us that the most widespread form, the M group of HIV-1, originated sometime between 1910 and 1930, and expanded rapidly in the 1950s, though the original zoonotic event (the jump from chimpanzees into humans) may have occurred earlier. Other introductions of less common forms of HIV-1 occurred independently, in some cases much earlier than the M group, but these strains also seem to have increased in prevalence at the same time. What was going on in the early 20th century to cause the current pandemic?

To think about this problem, we first have to understand how pathogens jump from a non-human host to a human (zoonosis, see Chapter 1). Zoonoses follow two typical patterns. A zoonosis with $R_0 < 1$ in the human population, such that each infected human infects (on average) less than one additional person, is called a spillover event. Individual infected people may become severely ill or die, but the disease won't establish in the human population because the transmission between humans is too low. On the other hand, if R_0 is even slightly greater than 1, the minimum value for persistence, then the pathogen can successfully shift hosts. Once the pathogen starts to circulate among humans, mutations can arise that help the pathogen adapt to its new host. In other words, these mutations open the compatibility filter further and further.

The more pathogens circulate in a population, the greater the number of genomes available to mutate and the greater the overall probability of compatibility-enhancing mutations. It is therefore critical to respond quickly to initial zoonotic events in order to keep the number of human infections, and thus the total mutation rate, small. For example, in 2003 an outbreak of Mpox virus (formerly known as monkeypox virus)—the first in the western hemisphere—occurred in the USA as a result of a two-step spillover from a shipment of infected Gambian pouched rats from Africa, to prairie dogs reared in captivity for the pet trade, to humans. However, because the patients were quickly given medical care, and because of good hygiene in the treatment facilities, human-to-human transmission failed to evolve in the USA and the epidemic was stopped after just 71 cases, with no fatalities. In Africa, human-to-human transmission of Mpox has evolved at least once (most likely several times). As we write this second edition, a new strain of Mpox has emerged which not only can transmit from one human to another, but has also succeeded in spreading to the Americas, Europe, and Australia. Epidemiologists who are old enough to remember the emergence of HIV in high income countries have noted disturbing echoes of 'an African disease that infects mostly men who have sex with men' (and hence risks being dismissed, once again, as someone else's problem).

The simplest argument for the current HIV pandemic, of course, is that nothing has changed—we have just been unlucky. According to this argument, the probability of SIV entering a human (passing the encounter filter) and evolving the capacity for transmission between humans (opening the compatibility filter) has been constant over time; the early 20th century just happened to be the one time it was successful. This argument is unsatisfying. Even if the probability of a successful host shift happening at any particular time is low, SIV, its primate hosts, and humans have coexisted for so long that it seems unlikely that such chimpanzee-to-human transmission has never happened before.

The most widely accepted hypothesis for HIV emergence is that HIV did indeed successfully jump from primates to humans many times, but that these events went unnoticed by the rest of the world because they failed to spread beyond small, remote villages. (A similar process has happened with the Ebola virus: small outbreaks have probably been going on in isolated human populations for centuries or millennia, and have been recorded over the past 40 years. It only caught the serious attention of the West with its emergence in the urban populations of West Africa in 2014.) However, the most recent HIV shift to humans coincided with a colonial context that facilitated transmission. Ongoing urbanization, population growth, and extremely high rates of sexually transmitted disease (as seen in documented rates of syphilis among European colonizers) all contributed to the emergence of HIV. A few decades later, changing patterns of global mobility, and social changes such as a gender revolution within Kinshasa as well as the sexual revolution in the USA and Europe, helped spread HIV across the rest of the globe.

Other researchers have argued that sexual transmission alone was not enough to spark the HIV pandemic. Colonial doctors reused non-sterilized syringes for vaccination, blood transfusion, and to treat sleeping sickness; these practices certainly amplified HIV spread, and may have facilitated the process of viral adaptation to humans.

These two hypotheses—urbanization and unsterile injections—are not mutually exclusive and may well have worked in concert to spread the virus. We will never know for certain, but by asking where HIV came from we have learned important lessons that we are already using to restrict the spread of other pathogens.

Chapter 5
Cholera

Cholera has a foundational role in any discussion of infectious disease. Cholera gave rise to Koch's postulates, one of the pillars of epidemiology. These postulates state that identifying a particular organism as the causative agent of a disease requires that (1) the organism is found in diseased people but not healthy ones, (2) the organism can be isolated from diseased people and grown in culture, (3) the laboratory-cultured organism will cause disease in a new host, and (4) the organism can be re-isolated from the newly infected host. Cholera also led to the recognition that preventive measures should focus on providing clean water, which in the process closed the encounter filter for a vast number of other water-borne pathogens.

The bacterium *Vibrio cholerae* is the cause of the diarrhoeal disease known as cholera. There are many species in the genus *Vibrio*, so we abbreviate the genus name to '*V.*' and retain the species name, as in *V. cholerae* (as opposed to, say, *V. vulnificus*, a close relative which is sometimes found in raw shellfish such as oysters and can cause acute gastroenteritis). *V. cholerae* itself is a large and diverse species, divided into groups known as *serotypes*, meaning 'types based on serum', because samples are classified based on their reaction to antibodies isolated from blood (or serum). The antibodies are very specific, so will only react with the same serotype of cholera that was originally injected into the mammal to produce the antibodies. In other words, if a new sample reacts with antibodies from an animal injected with a known strain of O1, it is classified as an O1 type.

Two serotypes, O1 and O139, are responsible for most epidemics. However, even within these serotypes, cholera bacteria can only cause epidemics if they carry at least two special genes: one to overcome the compatibility filter, and one to open the encounter filter.

To become infected with cholera, you have to ingest a huge number of the bacteria—approximately one million of them. Stomach acid, in addition to digesting our food, protects us by killing all manner of living things that we accidentally eat or drink, including cholera. Food acts to neutralize stomach acid; on a completely empty stomach, 100 *trillion* bacteria are required to reliably produce infection in humans, 100,000 times more bacteria than when ingested with food. With that many bacteria, some are likely to survive the gauntlet of stomach acid to make it through to the small intestine.

Bacteria that survive the stomach must also be able to infect cells in the small intestine. To do this, cholera must possess a cluster of genes involved in creating the toxin co-regulated pilus, or TCP. TCP is an essential component of the compatibility factor— without it, no illness occurs. Many other genes enhance colonization abilities, but TCP is the only one that is absolutely essential. Remarkably, even after colonization, illness usually does not result unless the gene for cholera enterotoxin (CT) is expressed (i.e. turned on so that the bacterium starts producing the enterotoxin protein); that is, without CT, cholera is infectious but not virulent. Virulence and transmission are strongly correlated in cholera: CT is essential for between-host transmission of cholera (its exit strategy). CT disrupts water regulation in the intestine, producing a flood of diarrhoea that leads to dehydration and death for the host but also pushes cholera back out into the environment, allowing it to encounter new hosts.

The bright spot in all of these horrific symptoms is that treatment for cholera is remarkably straightforward. Just giving cholera victims a simple oral rehydration solution containing sugar and salt can drastically reduce death rates. Antibiotics are useful, but as a secondary defence. The most important benefit of antibiotics is that they shorten the infectious period, and thus can reduce the risk of transmission.

Cholera has played an important role in the history of epidemiology. John Snow's discovery that cholera was spread by a contagious agent, and localizing that agent to a particular water pump during a cholera outbreak in mid-19th-century London, is arguably the first case of epidemiology as systematic detective work.

Snow's insights continue to be useful today: closing the encounter filter by setting up water purification and water treatment plants is an excellent way to prevent cholera epidemics. However, water purification by boiling (a low tech solution that can be implemented in remote or lower income areas) requires enormous amounts of firewood, which can be dangerous to retrieve in politically unstable areas, and can lead to deforestation. Higher tech solutions such as treatment plants or water purification via chemicals are expensive and can be logistically difficult. One inexpensive, low tech, effective intervention is to filter water through four thicknesses of sari cloth, which can reduce infection rates by 50 per cent.

However, even with water treatment systems in place, infected people can spread cholera within their own households. Analysing the social network of cholera epidemics shows that infections spread much more rapidly within households than between them, most likely due to contamination of water or food by cholera carried on the hands of caretakers. Accordingly, public health agencies have promoted water storage vessels with narrow mouths and spigots (rather than traditional open-mouthed buckets) so that infected people and their caretakers cannot contaminate the household's supply by dipping their hands or clothes in the water. Such interventions have reduced cholera transmission rates by almost 40 per cent.

How does cholera evolve? In the Red Queen race against hosts, bacteria generally use complete costume changes less often than viruses do. Because bacteria mutate many times slower than viruses, they can't accumulate individual changes fast enough to hide effectively from our immune systems. Accordingly, instead of immune evasion, bacteria focus on resistance to host countermeasures.

Rather than switching their entire wardrobe, bacteria such as *V. cholerae* accessorize by gaining (and losing) accessories (countermeasures) according to the demands natural selection makes of them at any given time. If natural selection demands a feather boa, and one is available in the costume box (i.e. the local environment), any bacterium that picks it up will prosper. That is, the bacterium with the boa will have more descendants than the ones without it—natural selection in the usual sense. However, to a bacterium even a feather boa is heavy, so if it is not explicitly required for survival it is in the bacterium's best interest to drop it.

Obviously bacteria don't wear boas. Instead, they gain and lose genes. One way to do this is by acquiring mobile genetic elements from other bacteria. Another source of useful novel genes is from viruses that infect the bacteria. Both these mechanisms are referred to as lateral gene transfer (LGT for short), because they involve the movement of genes among organisms within generations (laterally), rather than between generations (vertically). Bacteria reproduce by dividing rather than having offspring as animals do, but the creation of new individuals isn't required for LGT, just contact between existing organisms.

Given that the biochemical machinery of bacteria differs significantly from ours (in contrast to viruses, which often use our own biochemistry), we can fight bacteria by poisoning them with antibiotic compounds borrowed from other organisms such as fungi (Chapter 1). One of our favourite examples of evolution in real time is antibiotic resistance in bacteria, which is usually accomplished by LGT, often via genes carried on plasmids (small circular pieces of DNA that are separate from the chromosome). Cholera is no exception. While antibiotics are not required to cure cholera, they do shorten the duration of the infectious period, and the symptoms. However, heavy use of antibiotics will select for antibiotic resistance, if resistance genes are present in the population.

Unlike in viruses, drug resistance in bacteria does not generally arise as a result of new mutations, for several reasons. As already mentioned, bacterial mutation rates are usually much lower than viral rates—this reduces bacteria's ability to evolve new drug resistance mechanisms. And bacteria are more likely to trade genes among themselves—that is, participate in LGT—than viruses are. Two of the most infamous cases of drug resistance in bacteria, methicillin-resistant *Staphylococcus aureus* (MRSA) and carbapenem-resistant enterobacteriaceae, are caused by genes for resistance to different antibiotics.

Cholera acquires resistance primarily by gaining a piece of DNA called SXT, which is similar to a plasmid but is linear rather than circular. SXT and the multi-antibiotic resistance it confers coincides with much of the spread of the ongoing seventh pandemic of cholera. Once a bacterium gains antibiotic resistance, that bacterium will be more successful than those without resistance for as long as antibiotics are in use. When antibiotic use ceases, however, resistance is often lost. Bacteria with big genomes take longer to replicate than bacteria with small ones, so bacteria containing lots of integrated antibiotic resistance genes take longer to replicate than bacteria without such genes. This *cost of resistance* can be a significant handicap in a rapidly growing population.

How is resistance lost? Just as DNA can be integrated into the genome, it can also pop out again. If that happens, the bacterium with the smaller genome may have an advantage. Alternatively, if the antibiotic resistance genes are on plasmids, rather than within the bacterial chromosome itself, the plasmids themselves may be lost at random through genetic drift if they do not give the bacteria any advantage. In other bacterial systems, the cost of resistance may be due to the metabolic actions of the resistance gene, which may make the organism expend a lot of energy pumping toxins out of its cells, or because the organism switches to a biochemical pathway that is immune to the effects of the antibiotic but less efficient at its metabolic role.

Even as cholera bacteria parasitize us, they too can be parasitized—by viruses. Remarkably, as previously mentioned, bacteria can acquire useful genes from their viral parasites—although perhaps this isn't all that remarkable when we think about it. Natural selection rewards organisms for being opportunistic, so in some circumstances a virus that gives something useful to its host is more likely to persist than one that doesn't. To understand what kind of circumstance will allow cholera to pick up genetic novelties from its viruses, we need to know a little bit more about the viruses that infect bacteria: the *phages*.

One common type of phage is the temperate phage, which can follow two strategies. The first is lysogeny: once a phage enters the cell, its DNA is incorporated into the genome of the bacterium. The phage's DNA is now called a *prophage*. The prophage is replicated along with the bacterium's own DNA, spreading to all the offspring of the original infected organism. It's a peaceable kingdom—unless the prophage gets a biochemical signal that tells it that it is threatened. In that case, the phage may exit non-destructively, or it may adopt a *lytic strategy*.

In the lytic strategy, the phage turns its bacterial host into a virus production factory, ultimately killing the bacterium by bursting ('lysing') the cell once enough offspring are created. These new phage can go on to infect other bacteria, and pursue either the lysogenic or lytic strategy depending on environmental conditions.

As mentioned earlier, only a few strains of *V. cholerae* carry the toxin gene CT. CT is a gift carried by the temperate phage CTXϕ to its host. Once CTXϕ is ensconced in the cholera genome, the toxin can be expressed and thus can increase transmission (and hence R_0). But phages, even temperate phages, are not altruistic. CT increases the fitness of the bacteria, but it also increases the fitness of the (now) prophage, which multiplies right along with the bacteria inside the patient's gut.

The other essential genetic ingredient for pandemic cholera is TCP, which enables cholera to colonize the small intestine. TCP has another function as well: it is the receptor for the phage CTXϕ. Receptors are part of the host compatibility filter for viruses. Without the correct match between virus and receptor, infection can't occur. By expressing TCP, cholera invites infection by CTXϕ, and thereby receives the gift of CT. It's a beautiful, if sinister, example of coevolution: the two components that are necessary for high transmission of cholera (and therefore also of the prophage) are part of an interdependent system assuring the fitness of both parties.

But this pretty scenario of invitations and gifts is a profound misreading of what is more accurately viewed as an ongoing feud. The bacteria do not want the phages to come to the party, gift or no; even if the phage doesn't kill when it exits, it still incurs a cost, however small, on the bacterium. Instead, the phage is a party crasher. It uses TCP as a receptor because TCP is something the bacterium needs for its own sake. And CT is not a gift. The phage carries it only because it increases its own fitness. If the phage's fitness can increase along with the bacterium's, fine, but the only reason it works that way is because the phage is integrated into the bacterial genome, and so their fitnesses are inextricably linked. But if a mutation or new chunk of DNA were to appear that increased phage fitness still further at the expense of the bacterium, or that allowed the bacterium to maintain CT while getting rid of the rest of the party-crashing prophage, natural selection would optimize such changes and all illusions of polite society would vanish.

There is some evidence for the darker version of this story. Digging a little deeper, researchers have learned that the genes coding for CT are probably a recent acquisition by the phage. There are many related phages that do not carry CT, but still use TCP as their receptor. Moreover, the composition of the CT genes suggests that the CT genes are newer than the rest of the genome. Not only bacteria, but also the viruses that parasitize them, can accessorize for their own benefit.

Why hasn't TCP evolved so that the CTX phage can't use it, particularly since many of these phage don't even bring CT with them? Phage infection probably doesn't hurt the cholera bacteria very much—even when threatened, this phage exits the cell without killing it. So there's little selective pressure for the bacteria to resist infection by altering TCP, especially given that they need TCP for intestinal colonization.

Other phages that infect cholera behave more like traditional predators which always kill their prey. These phages are *obligately* lytic; they can only survive by lysing or bursting their host. If they infect a bacterium, it's doomed. The best known of these lytic phages are the JSF phage group; they use a different receptor from CTXφ to gain entry to the cholera bacteria.

In regions where cholera is endemic, epidemics occur seasonally. Scientists have noticed that epidemics in the Ganges River Delta tend to start when there are lower numbers of lytic phages in the water and stop when there are more lytic phages. Moreover, the relative abundance of phage in patients with cholera reflects their relative abundance in the aquatic environment: fewer lytic phage in patients early in the epidemic, more later in the epidemic. These observations suggest that lytic phage epidemics among cholera might ride along on top of the cholera epidemic occurring in humans, possibly even helping to control the epidemic.

Phages could be used to control cholera in individuals as well as at the population level. This form of treatment—phage therapy—has been around at least since the 1920s, when it was developed as a treatment for dysentery, a diarrhoeal disease like cholera that is caused by a different bacterium. The big advantage of phage therapy is that many phages are specific to a particular bacterium, so there are no side effects. Conventional antibiotics attack not only unfriendly bacteria but also the essential bacteria in our bodies; many women have had the experience of suffering from yeast infections when antibiotics have knocked out the bacteria that normally inhabit their reproductive tracts. Successful, well-designed phage therapy would kill the bad bacteria while leaving the good guys alone.

The downside of phage therapy is that it tries to harness a living system that is capable of evolution. Once phages are introduced to a patient, they can evolve in any way natural selection directs them, including in ways that could make people sicker, as in the case of CT. One possible safeguard that could reduce such risks of phage therapy is to limit therapeutic phages' capability to exchange genes with the bacteria they infect. Phage evolvability may be one reason that phage therapy has not been embraced outside Russia and Georgia, though very few adverse incidents involving phage therapy have been reported in the last 60 years. Phage therapy is enjoying a resurgence of interest around the world, in part due to the rise in antibiotic resistance, but also in part due to the apparently beneficial effects of lytic phages at the population level.

The seasonality of cholera epidemics is worth a closer look. In Bangladesh, phages parasitizing cholera fluctuate seasonally just as they do in India. These fluctuations seem to track the density of cholera in a pattern that is well known to ecologists as the pattern of regular (animal–animal or animal–plant) predator–prey systems: the prey (cholera) increase first, then predators (phage) increase, causing prey to decrease once again. Once prey decrease, the predator of course decreases as well, having nothing to eat, allowing prey to increase once again and creating cycles of abundance. Thus, rather than the phage controlling the abundance of their prey, *V. cholerae*, the prey might well drive the abundance of the predator. Cholera epidemics in Haiti also show seasonality, but there don't seem to be any phages in the environment. Thus, either seasonality is driven by different mechanisms in the two locations, or the phage seasonality is a consequence rather than a cause of bacterial seasonality.

Contrary to what one might expect of an organism transmitted through drinking water, cholera thrives in salt water. It can also persist in fresh water, if the water is warm enough and contains enough nutrients. Epidemics in Bangladesh closely track surface seawater temperatures, for two reasons. First, warm sea surface water temperatures are associated with severe storms, which in turn cause flooding. Flooding worsens existing sanitation problems, co-mingling drinking water and sewage, and facilitating transmission of cholera.

Second, warm sea surface temperatures (particularly in association with large amounts of nutrients) cause blooms of tiny plants called phytoplankton. The phytoplankton are food for zooplankton—tiny floating marine animals. Cholera can stick to the exoskeletons of shrimp-like zooplankton known as copepods, whose bodies are made of a substance called chitin. Remember that a person needs to swallow huge numbers of cholera bacteria in order for any of them to survive our stomach acid. An individual copepod can accumulate 10,000 cholera bacteria: swallowing such a copepod is like ingesting a cholera bomb. If one such copepod arrives on a stomach full of food, enough cholera can easily be delivered to cause disease, while swallowing water with just a few free-living bacteria is unlikely to harm anyone. The ability of large numbers of cholera to stick to copepods has been used to explain how cholera outbreaks could arise from environmental sources of cholera, rather than from infected patients.

As with most areas of science that matter at all, some parts of the cholera story are still contentious. Scientists have long tried to understand how cholera can spread through a population so terribly quickly—for example, there were 30,000 cases in just the first week of the 1991 cholera epidemic in Peru. Some researchers attribute these explosive outbreaks to contamination of key water supplies and seasonal changes in the environment. Others point to the puzzling fact that cholera behaves differently depending on the conditions under which it was reared. Cholera reared in laboratory culture are the standard on which estimates of the infectious dose (the number of particles required to make a person ill) are generally calculated. However, when lab-reared cholera fed to a volunteer go through the process of infecting the gut and being expelled in diarrhoea, they can change to a *hyperinfectious* state. Far fewer numbers of these hyperinfectious bacteria are required to infect a new host.

If hyperinfectious cholera reaches the gut of another human, it remains in the hyperinfectious state (as long as it doesn't get killed by lytic phage or hit with antibiotics). On the other hand, if it leaves a patient's bowels, enters the water, and fails to get taken up by another human victim within 5 to 24 hours, it undergoes another fundamental physiological change to a very different state called either 'viable but non-culturable' or 'active but non-culturable'. The non-culturable form of cholera has gained attention as a potential environmental reservoir for new cholera outbreaks. However, it's unknown whether or not this process is reversible, in particular, whether or not the bacteria can recover their ability to colonize humans after entering the non-culturable state.

One example of the importance of understanding sources of cholera outbreaks and the virulence of genes themselves comes from the recent (2010–19) cholera epidemic in Haiti. The tragedy of the 2010 earthquake in Haiti was heightened by an outbreak of cholera, with over 470,000 cases reported and 6,631 people dead in the first year. Haiti had been free of cholera cases for at least 100 years, which raised the immediate question of how the disease had arrived. Analyses ranging from serotyping, to simple comparison of presence and absence of particular genes, to sophisticated phylogenetic analyses, have all suggested that the strains did not come from a local environmental source, such as the Gulf of Mexico.

Instead, the strains isolated from Haitian patients were most similar to strains from South East Asia, suggesting that they may have arrived with people from this area, or people who recently visited there—in this case, UN peacekeepers from Nepal. In addition, the camp of the peacekeepers was located on a tributary of the Artibonite River, which was identified as the source of the Haitian outbreak, although cholera rapidly spread throughout the country.

A safe, effective (i.e. how well the vaccine works in ideal conditions), and efficacious (in other words, how well the vaccine works in real life conditions) oral vaccine for cholera exists. So why isn't vaccination routinely emphasized in the case of cholera, given the large toll of the disease? In outbreak situations and/or humanitarian crises expected to result in outbreaks, vaccination is indeed recommended. However, in non-outbreak situations, WHO recommends use of these vaccines only as a complement to, rather than a replacement for, other strategies such as water purification systems and handwashing campaigns. The reason is that infrastructure and behavioural changes are sustainable, long-term solutions to the problem of cholera; vaccination is not, as the efficacy of the vaccine wanes after a few years (or, in the case of only a single dose, after six months). Moreover, water, sanitation, and hygiene (WaSH) infrastructure protects the population not only from cholera, but from many other water-borne parasites as well.

These ideal infrastructure solutions often cost much more in the short run than vaccination, leading to capital-based tradeoffs between sustainability and immediacy. This problem applies particularly to countries with limited resources, the very places where cholera is most common. However, the investment needs to be considered in light of the economic burden of cholera. Such a burden includes strain on the health system resources due to care of cholera patients; lost productivity by patient caregivers, as well as patients themselves; lost productivity due to premature death; etc. In Asia, the economic burden of cholera was estimated to be a minimum of 10 per cent GDP per capita per day for 2015. Such numbers put the potential return on investment for water safety infrastructure in sharp focus and underscore the need for international cooperation in these efforts.

Chapter 6
Malaria

Malaria might be the most important infectious disease on the planet. Compared to the infectious diseases discussed in the previous chapters, it is less frightening to people in temperate, high income countries—not because it is less infectious or less virulent, but because in modern times it rarely reaches out of the tropics, being limited by the ecological niche of its mosquito vectors. Unlike cholera, malaria tends to be *endemic*—the number of cases is fairly constant across years, with a strong seasonal pattern—rather than occurring in intense epidemic outbreaks. Typical of endemic disease, the most widespread species of malaria are chronic and debilitating, rather than causing acute infection and death. The exception is falciparum malaria, most common in tropical sub-Saharan Africa, where it is one of the most common causes of death. The fever, malaise, malnutrition, and anaemia associated with chronic malaria are associated with poorer educational outcomes in children, while acute malaria can lead to chronic neurological problems. Combining these non-lethal effects common to all malaria species with the lethal effects of falciparum, the cumulative impact of malaria on humanity is enormous.

Public health officials measure the impact of chronic diseases in terms of *disability-adjusted life years* (DALYs), which take into account both the loss of life and the loss of productivity and happiness due to a disease. Coming up with appropriate weights poses obvious ethical challenges. Is one year of life with a chronic, crippling disease equivalent to six months of healthy life? Three months? One month? How should one compare the impact of death or disability in an infant, a middle-aged person, or an elderly person? Nevertheless, DALY calculations account for the fact that the chronic effects of disease on people living with disease can be just as important, if not more important, than disease-induced deaths. In 2010, malaria accounted for more lost DALYs than any other specific infectious disease (not counting the broad classes of 'lower respiratory infections' and 'diarrhoeal diseases', which kill young children and so account for many years of lost life), narrowly edging out HIV/AIDS; as of 2019, thanks to control efforts, it had fallen to third place (after HIV/AIDS and tuberculosis). Although the evidence is hard to disentangle—poor countries tend to be malarial while rich ones do not, and malaria affects economics and society in many ways—researchers have suggested that eliminating malaria could increase economic growth rates by several percentage points, the difference between a struggling economy and a healthy one.

Although every infectious disease has its intriguing quirks, malaria is more complex than the other diseases we have discussed so far. It is caused by a protozoan, a single-celled organism whose genetic material (unlike in bacteria) is confined within a nucleus. Its genome comprises about 23 million base pairs of DNA, thousands of times larger than the viral genomes of HIV and influenza and about five times larger than cholera's genome.

Malaria's complexity also stems from its *vector-borne* nature; it is transmitted from one human to another by various species of *Anopheles* mosquitoes. Cholera adjusts its biochemistry to function alternately as a free-living organism in the ocean and as a pathogen in the human gut. Malaria has an even more challenging problem: it adjusts its biochemistry to two different biological environments (the human host and mosquito vector), and to multiple organs within each host (human blood and liver; mosquito gut and salivary glands). Biological environments represent bigger challenges for parasitic organisms than physical environments. Desiccation and ultraviolet light are dangerous threats to parasites travelling through the physical environment outside the host, but they are passive. In contrast, host organisms take active steps to destroy parasitic hitchhikers by attacking them with immune defences.

Four species of malaria commonly infect humans. In order of decreasing severity, the malaria species are *Plasmodium falciparum*, *P. vivax*, *P. ovale*, and *P. malariae*. (As is the case with cholera, malaria species are referred to by the genus name *Plasmodium*, which can be abbreviated as *P.*, and their species name.) A fifth species, *P. knowlesi*, known originally as a disease of macaques, is emerging as a disease of humans in South East Asia, but so far human–mosquito–human transmission, as opposed to monkey–mosquito–human transmission, is rare.

All malaria species have essentially the same life cycle. It starts with sexual reproduction (fusion of female and male *gametocytes*, the equivalent of eggs and sperm coming together) in the gut of a mosquito host, after which the next life stage (*sporozoites*) migrate to the mosquito salivary glands and are injected into the human host when the mosquito bites and sucks blood to get protein to feed her eggs. (Only mature female mosquitoes bite humans, which has important implications for malaria control.) The injected sporozoites migrate to the human's liver where they reproduce by simple division, migrate back to the bloodstream, and continue to multiply, infecting and destroying red blood cells as the population grows. Eventually the blood stages develop into female and male gametocytes and wait for another female mosquito to arrive and suck them up.

Malaria causes anaemia by destroying red blood cells; it also contributes to low birth weight in infants, causing numerous health problems, when malaria-infected red blood cells infect the placenta. However, the worst symptoms occur when malaria parasites (most often *P. falciparum*) get into the brain, causing cerebral malaria. In cerebral malaria, parasites block blood flow and trigger inflammation, killing untreated patients and often causing brain damage even in patients who are treated and recover.

Malaria's most obvious symptom is fever, although fever does not itself seem to be harmful. Malarial fever recurs with characteristic frequency as new waves of parasites emerge from the liver into the blood: historically malaria strains were classified according to whether fever recurred every other day (*tertian*: *P. vivax* and *P. ovale*) or every three days (*quartan*: *P. malariae* and *P. falciparum*). In many malarial regions, patients who come to the hospital with fever and headache are automatically treated for malaria since testing is expensive and requires expertise.

Malaria's inability to move directly from one human to another opens up a wide range of possibilities for control. As Chapter 2 describes, Ronald Ross, one of the founders of disease modelling, figured out two important facts about malaria. First, he discovered that mosquitoes transmit malaria, which focused attention on preventing malaria by controlling mosquitoes, or by preventing them from biting humans (closing the encounter filter). Second, his models showed that public health agencies didn't need to completely eradicate mosquitoes to eradicate malaria—they just needed to reduce the number of mosquito bites by killing or repelling mosquitoes, until, on average, each malaria-infected human is bitten by so few mosquitoes that they lead to fewer than one new human infection. Stopping mosquito bites is especially effective because a mosquito must take two bites to complete the infection cycle—the first to get infected, the second to transmit infection to a new human host. The encounter filter doesn't need to be completely shut—just closed tightly enough that only a few mosquitoes can sneak through.

If you can't close the encounter filter by killing, blocking, or repelling mosquitoes, you can try to close the compatibility filter. In prehistoric times, humans evolved many genetic mechanisms for closing the compatibility filter, though always at a cost. Even before humans knew the cause of malaria, we developed drugs to block the compatibility filter by poisoning malaria within our bodies in ways that are somewhat less toxic to humans than to malaria. For example, gin and tonic was the favoured drink of British colonists in the tropics from the early 1800s on, due to the antimalarial action of the quinine found in tonic water (modern tonic water has much lower concentrations of quinine, which are less toxic for the general public than medicinal doses). Most recently, we have developed vaccines to bolster our natural immunity, although so far without complete success. In what follows we will discuss all three of these compatibility-blocking strategies.

Biologists have retrieved ancient DNA from 4,000-year-old Egyptian mummies, but we know malaria is far older. Early relatives of malaria have been found in the guts of midges (biting flies) in amber, which were probably biting cold-blooded reptiles, from 100 million years ago. However, such fossils are extremely rare, and in order to resolve the gap between 4,000 and 100 million years ago we have to turn to the genome of malaria, and of its hosts.

Malaria is part of a large, complex family of parasites, the apicomplexans, that frequently jump between hosts. Although the malaria parasites of lizards, birds, and mammals are all called *Plasmodium*, they are most likely two separate families (one that infects lizards and birds and one that infects mammals) that are more closely related to other parasites than to each other; mammalian malaria split off from the rest of the family around 13 million years ago. The malaria species that infect humans are a similarly mixed group; they are all more closely related to various malaria parasites of non-human hosts than to each other, having split off somewhere around two to seven million years ago.

Establishing the origin of human malaria species is challenging. Analyses of the genome of *P. falciparum*, the most dangerous species of human malaria, suggest that its population expanded greatly about 10,000 years ago, around the time of the development of agriculture and a concomitant increase in human population density. Where *P. falciparum* came from in the first place is a tougher question. It is closely related to the chimpanzee malaria *P. reichenowi*, leading older textbooks to state that it jumped from chimpanzees into humans. However, over the past decade extensive sampling of the faeces of wild primates in tropical sub-Saharan Africa, combined with advances in genetic technology that allow researchers to sequence the genomes of individual malaria parasites, have established that *P. falciparum* is most closely related to a malaria species of gorillas. (Earlier detections of *P. falciparum* in captive bonobos are now thought to represent an instance of 'spill-back', where wild animals are infected by human parasites.)

The human genome sheds further light on the history of malaria. Humans are not passive nurseries for malaria parasites; our immune systems are constantly evolving new ways to counter the parasite's debilitating effects, although always at a cost to ourselves. Humans have evolved many genes that fight malaria, with varying efficacy and severity of side effects. The best known of these strategies is the sickle-cell trait, which appears in biology textbooks as an example of heterozygote advantage: individuals with one copy of the sickle-cell allele and one normal haemoglobin allele are tolerant of falciparum malaria, but having two copies of the sickle-cell allele causes crippling anaemia.

Thalassemia is a disease similar to sickle-cell anaemia, most common in Mediterranean populations, caused by a gene variant that modifies haemoglobin production in a way that reduces the severity of malarial symptoms at the price of anaemia. Glucose-6-phosphate-dehydrogenase (G6PD) deficiency, also common in Mediterranean populations, protects against falciparum and vivax malaria. G6PD also causes anaemia, but only under particular circumstances such as eating fava beans or taking common antimalarial drugs such as chloroquine or primaquine. Duffy negativity—the absence of the 'Duffy antigen', a protein that *P. vivax* and *P. knowlesi* target to enter red blood cells—prevents against malaria symptoms at the cost of a wide range of side effects.

Genetic analysis can estimate how long ago mutations arose. The well-known human blood type O appears to provide some protection from malaria, but it is so old—it has been around for millions of years, longer than human malaria itself—that it must originally have evolved for some reason other than malaria protection, probably to protect against some other now-unknown blood pathogen. Some variants of Duffy negativity are around 30,000 years old. G6PD deficiency arose 5,000 to 10,000 years ago, reinforcing the evidence from the falciparum genome that the risk to humans from falciparum malaria exploded around the time that agriculture developed. In contrast some sickle-cell variants are evolutionarily young—only a few hundreds or thousands of years old—reminding us that the human and malaria genomes are constantly (co)evolving.

Because these genetic protective mechanisms come with severe side effects, natural selection increases their frequencies only in regions where the risk of malaria outweighs the side effects. Elsewhere, it reduces their frequencies. Since malaria is so widespread and has such deadly effects, it has been one of the strongest selective forces shaping the human genome in the past few thousand years. The geographic distributions of particular malaria-related genes (the Duffy antigen and sickle-cell in sub-Saharan Africa, thalassemia and G6PD deficiency around the Mediterranean) give evidence about the historical distribution of malaria. Because malaria and malaria-protective genes are only two of the vast number of factors affecting human fitness, however, the stories told by human genetics and pathogen genes are sometimes complex. For example, researchers have traditionally thought that *P. vivax* originated in sub-Saharan Africa, since humans there often carry mutations that deactivate the Duffy antigen (which in turn protects them against *P. vivax* infection). Studies of the *P. vivax* genome complicated the story by linking *P. vivax* most closely to macaque (monkey) malarias from South East Asia, suggesting that Duffy negativity might have evolved for protection against other species of malaria or other malaria-like parasites, or after *P. vivax* made its way from Asia into Africa. Most recently, however, malaria DNA from wild gorilla faeces has again shifted the evidence back in favour of an African origin for *P. vivax*.

In addition to humans' evolved constitutive defences against malaria (systems that are in place whether or not a person has ever been infected), humans' adaptive immune systems can also help. Unfortunately, unlike the immune response to simple, acute viral diseases like measles, where immunity develops quickly and is essentially lifelong, malaria immunity develops slowly. No one completely understands why, although it probably has to do with the genetic diversity of malaria, and the ability of a single clone of malaria to switch its molecular appearance. In high malaria areas, the frequency and severity of malaria symptoms declines with exposure. Malaria immunity generally wears off quickly, perhaps in part through the malaria parasite's interference with the human immune system, so that malaria is a lifelong threat, though it is less common in older children and adults.

Another problem is that humans' adaptive immunity to malaria provides more tolerance (also called *clinical immunity*) than resistance. In other words, immunity does reduce the number of parasites in the bloodstream, but its main effect is to reduce the severity of symptoms. This tolerance has two important implications: (1) people in high malaria zones with clinical immunity don't feel sick when they get infected, so they won't get treated even when treatments are available—this makes it harder to reduce malaria rates from high levels; (2) it is correspondingly easier to keep malaria rates in check once malaria rates, and levels of clinical immunity, have been reduced.

For most of history, humans have just had to live with malaria. Our historical knowledge of malaria begins with humans' first attempts to consciously defend themselves against malaria. Humans discovered compatibility-blocking chemical defences against malaria long before they discovered the microorganisms that cause it. Jesuit priests brought the traditional medication quinine, derived from tree bark, to Europe in the early 17th century. Because South American natives chewed on the bark of the cinchona tree to stop shivering, the Jesuits guessed it might work on malarial fevers, which are often accompanied by shivering. The Jesuits got lucky; quinine doesn't actually reduce fever, but it does cure malaria by leading to the build-up of toxic chemicals within blood-inhabiting stages of the malaria parasite. (Why were Jesuits getting malaria in South America, when malaria evolved in sub-Saharan Africa? *P. falciparum* and the milder *P. vivax* were both brought to the New World by the trade in enslaved African people, although it seems most likely that other strains of *P. vivax* were already circulating there, brought earlier by ocean voyagers from South East Asia.)

While quinine can cure malaria, it is too expensive and too toxic to give it to everyone in the population regardless of whether they have malaria or not. It may be cheaper and safer to prevent disease by closing the encounter filter to block transmission rather than trying to close the compatibility filter once transmission has already happened. This idea applies to many diseases. Lower-tech solutions like monogamy, or condoms, or clean needles, or hand sanitizer, or insecticides, can work better (if you can get people to use them) than the best vaccines and treatments, and are especially effective when used in concert with these compatibility-blocking measures.

For malaria, the conclusion from this line of reasoning is that controlling malaria transmission (e.g. by adding *larvicides*, chemicals or specialized bacteria that kill juvenile mosquitoes, to potential mosquito habitat) may work better than treating people who are already infected. Sometimes mosquito reduction, and thus malaria control, occurs naturally as a side effect of changes in land use; changes in agricultural practice that reduced the amount of standing water available for mosquito breeding are thought to have led to declines in malaria in the northern USA in the late 19th century. Of course, land use change can work in either direction. Abandonment of agricultural land in the southern USA in the 1930s *increased* malaria infection. More recently, malaria in western Kenya has increased along with an increase in the number of active brick-making pits, which hold water but tend to have few mosquito-eating predators, providing a perfect habitat for larvae; deforestation is associated with increases in malaria in the Amazon, although this may be due to the settlement of humans in forest edges rather than with landscape modification per se.

Prior to the discovery of DDT, public health authorities sometimes teamed up with engineers to reduce larval mosquito habitat by managing water—draining swamps or increasing water flow or changing water levels to make the habitat inhospitable to whatever local species of mosquito was transmitting malaria. They also sprayed oil or arsenic-based insecticides in the water to kill larvae. In a form of biological control, health authorities introduced larvae-eating fish, especially the genus *Gambusia*, which is called the mosquitofish due to its dietary habits. In the first half of the 20th century, mosquitofish were brought from

North America to malarial regions all over the world, from South America to central Asia to Italy to Palestine. However, it's hard to know exactly how well they worked since they were usually combined with other control measures such as water management and chemical spraying.

While such forms of source reduction were somewhat effective in higher income countries, temperate or semi-arid regions, and other areas where malaria spread slowly, they were often impractical for poor, humid, tropical countries where malaria was rampant. Paul Hermann Müller's discovery of the insecticidal properties of DDT in 1939, for which he won the Nobel Prize in 1948, revolutionized malaria control. It triggered a shift from environmental and ecological engineering focused on destroying mosquitoes in their larval habitats, to protection by killing adult mosquitoes in the vicinity of humans. Indoor residual spraying programmes apply long-lasting insecticide to surfaces in houses where mosquitoes like to land. They work extremely well if a cheap, long-lasting, relatively non-toxic (to humans) insecticide such as DDT is available. Houses may only need to be sprayed a few times a year, depending on the climate and the characteristics of the wall surfaces. In general, such *residual spraying* works better than trying to exclude mosquitoes from houses—not only is it usually cheaper, but it kills mosquitoes rather than just excluding them, thus potentially providing some protection in the area around houses. Residual spraying programmes recorded initial successes all over the world: malaria was finally eliminated from the USA in the early 1950s after decades of source reduction, while residual spraying (with DDT and with other insecticides) in many African countries significantly lowered the malaria burden, at least initially.

The second great post-war advance in malaria control was the development of chloroquine, a cheaper and relatively non-toxic biochemical variant of quinine. Chloroquine was first synthesized by German scientists after the First World War. At that time the

Allies controlled Java, which had most of the world's supply of quinine, depriving German troops in East Africa of this vital drug. During the Second World War, the tables turned when the Japanese took over Java; US researchers then developed chloroquine into an effective antimalarial, although again too late to help their soldiers (who were exposed to malaria in Sicily and South East Asia). The mass administration of chloroquine both improved individual health by curing sick people of malaria, and reduced the threat of malaria at the population level by reducing the chances that a malarial person would pass the pathogen on to a mosquito. In addition to its other advantages, chloroquine helps control the sexual stages of most malaria species, thus blocking transmission as well as helping the infected person.

The combined power of DDT and chloroquine, along with other synthetic insecticides and treatments, raised the hopes of global health agencies in the 1950s that malaria could be controlled and even eradicated. However, malaria control efforts in the most resistant zones—rural districts of poor, tropical countries—ran into unanticipated problems after a few decades. Even though DDT and chloroquine were relatively cheap, non-toxic, and effective, the financial and logistical difficulties of mounting a global malaria campaign were greater than anyone had imagined.

First, like all campaigns to help people in middle and low income countries, malaria control programmes run into logistical, administrative, and cultural roadblocks. Does the local government actually send the supplies out to the places they're most needed? Do people steal them for other uses? Are the roads good enough to get them there? If they get there, can you train people to use them properly? Local residents may not want to use DDT, because it smells bad and stains their walls. They may resent foreign intrusion. Conflicts may break out and interrupt the programme, sending you back to square one. Funding agencies may run out of money after a few years, or they may decide that some other problem—feeding people, or providing them with clean water, or preventing a different disease such as Ebola—is more important.

The Global Malaria Eradication Programme, which began in 1955 and aimed to eradicate malaria by 1963, encountered all of these problems, and more, before it was finally abandoned in 1969, although it did reduce the malaria mortality rate more than 10-fold from its 1900 baseline. Another unforeseen problem with the campaign was the linkage of DDT use to bird deaths in North America publicized in Rachel Carson's *Silent Spring* (1962), leading to a ban on the pesticide within the USA and later in many other countries, although its use for vector control is still condoned by global environmental treaties.

Second, like all infectious disease control efforts, the enemies you're fighting are biological organisms that evolve countermeasures against your control strategies. Resistance evolves both in malaria vectors (against DDT and other synthetic insecticides) and in the pathogen itself (against chloroquine and other synthetic antimalarial drugs). The basic principle of chemical control is to poison the target organism—introduce a chemical that disrupts some aspect of its physiology or biochemistry, without being too toxic to the host (for chemotherapeutic agents) or other species in the environment (for vector control agents). The target is then under strong selection pressure to either (1) change its biology in a way that neutralizes the effects of the chemical, or (2) develop ways to detoxify the chemical or remove it from its cells.

Mosquitoes gain resistance to chloroquine by inheriting mutations that pump the chemical out of their cells. No one knows exactly when these mutations first occurred, but they spread very rapidly; chloroquine resistance in *P. falciparum* was first detected in the late 1950s in South America and South East Asia, and appeared in Africa in the 1970s. By the mid-2000s, chloroquine resistance had spread globally. Luckily, a new antimalarial drug called artemisinin, rediscovered by Chinese researchers screening historically known cures for fever, was already available. Artemisinin is now the first-line drug of choice for malaria treatment, in combination with other compounds. However, it, too, has already seen the evolution of partial resistance in South East Asia, and resistance is beginning to emerge in Africa as well. The World Health Organization is actively trying to prevent resistance from spreading through a crash malaria control programme in the Mekong region of South East Asia where artemisinin resistance is most prevalent; despite disruptions due to the COVID-19 pandemic, this programme has succeeded in drastically reducing falciparum malaria, aiming to eliminate it from the region by 2023. In areas where resistance has not yet emerged, public health agencies try to make sure that artemisinin is always used in combination with other antimalarial drugs to decrease the probability that strains of malaria will be able to evolve to resist both drugs simultaneously, similar to the design of antiretroviral treatment for HIV.

DDT-resistant mosquitoes have genetic modifications that either change the biochemistry of their neurons (the target of DDT), or allow them to detoxify DDT within their bodies. DDT resistance had already begun to be detected by the mid-1950s, and is widespread today, although levels of resistance vary enormously from country to country and from one mosquito species to another. Some researchers argue that DDT resistance comes in large part from heavy agricultural use (against insects other than mosquitoes, but exposing mosquitoes as a by-product), and that DDT could have been much more effective if it had been restricted to use in disease control programmes. While in principle mosquito populations could also lose DDT resistance if managers stopped using DDT in an area, experiments in flies have shown that DDT resistance does not seem to lower their fecundity, so DDT resistance is likely to persist for a long time. Malaria control programmes now try to manage resistance by switching among a variety of different synthetic insecticides that vary in their cost, effectiveness against mosquitoes, and toxicity to humans or other insects or wildlife.

In the last few decades international efforts towards malaria eradication and control have ramped up again. Planners have learned lessons from the failures of earlier programmes: in particular, they are more aware of the importance of political and cultural context, of using different combinations of strategies in different regions, and of considering malaria control programmes as part of a more general improvement in health infrastructure. They have also scaled back their optimism about eradication: the new international programme is called the Global Malaria *Action* (rather than 'Eradication') Plan (GMAP), and aims only for local elimination from particular countries. While GMAP does state that spending on malaria control can decrease as malaria is eliminated from some countries, so that money only needs to be spent making sure that it stays eliminated, only about US$4 billion per year (as of 2019) is being provided compared to the US$6–10 billion that is most likely required for control and eradication—and this level of commitment would probably need to be maintained for decades.

Current efforts to control malaria have two new tools that were not available in the 1960s. The first is insecticide-treated bed nets (ITNs) using pyrethroids, a class of insecticide that is safe for mammals (although toxic to fish). Pyrethroid-based ITNs were first deployed in the 1980s; long-lasting versions that can kill mosquitoes for five years beyond the six months' effectiveness of the original ITNs were deployed in the early 2000s. ITNs work similarly to indoor residual spraying—they kill mosquitoes that come indoors to bite humans—but they have the additional advantages of providing a physical as well as a chemical barrier, and working even in houses that have porous walls unsuitable for spraying. They are not without logistical and cultural problems—for example, recipients have been known to use the nets for catching or drying fish rather than for malaria protection. However, given that no malaria control strategy is 100 per cent effective by itself, ITNs provide a vital addition to the malaria control arsenal.

One lesson of bed nets is that a long-lasting tool is nearly always better—more cost-effective and requiring less effort to deliver—than its short-lived counterpart. A bed net that needs to be replaced every five years is better, all other things equal, than a residual spraying programme that needs to be repeated every six months, even though the bed net only works when people sleep under it.

An even newer technological innovation with long-lasting effects is the recently developed RTS,S vaccine, which has recently been deployed in a large-scale pilot study of 10 million doses for children in Ghana, Kenya, and Malawi. Compared with classic vaccines like those for smallpox and measles, which are more than 90 per cent effective, the new vaccine has 'modest' efficacy—typically below 50 per cent, depending on time since vaccination, age group, and which outcome is being prevented (e.g. 'clinical' malaria, i.e. a malaria case requiring a visit to a doctor or clinic, or severe malaria, involving seizures or other life-threatening symptoms). Furthermore, the vaccine requires multiple shots and provides immunity lasting only a few years, and it is less cost-effective (measured in 'dollars per case averted') than ITNs. Although it does prevent onward transmission of malaria by targeting malaria protozoans before they get into the host's bloodstream, its main use is to alleviate symptoms and save lives rather than controlling overall transmission. Nevertheless, in areas with high malaria transmission, it is still a valuable addition to the arsenal of malaria control strategies. Other, possibly more effective, vaccines are under development but have different challenges such as the need for extremely low-temperature storage.

We are still deeply ignorant about malarial parasites. Knowing more about the ancient history of different malaria strains—when, and from where, they entered the human population—would improve our understanding of the zoonotic processes that give rise to new human strains. Knowing more about the current distribution and ecology of malaria in non-human primates could help us guess about the likelihood of future zoonotics. Increased sampling of wild populations, and faster and cheaper genomic scans of malaria and primate genomes, are helping to resolve the picture, but it is anyone's guess how completely we will ever understand either the ecology or the evolutionary history of malaria.

So what does the future hold for malaria control? There is some room for optimism. In the absence of major economic or political shocks to tropical regions, or the continued emergence of pathogens that shift the focus away from malaria, the current efforts of many countries and foundations will continue to chip away at the burden of malaria over the next few decades. Elimination is possible in many areas, but malaria experts are at best cautiously optimistic about eradication.

Along with the evolution of mosquitoes and malaria, climate change and (more importantly) land use and economic change will continually move the target. No single magic bullet will solve the problem of malaria. Governments and agencies will have to deploy different combinations of the available tools (source control, residual spraying, bed nets, drugs for both prevention and treatment, and eventually vaccine) in ways that are appropriate to the local situation.

Chapter 7
Amphibian chytrid fungus

Amphibian chytrid fungus (*Batrachochytrium dendrobatidis*, or 'Bd' for short) differs in many ways from the earlier examples. This fungus is the first (and only) non-human pathogen we will discuss. Its genome is many times larger than the viruses and bacteria considered earlier, similar in size to the malaria protozoan's. It is a generalist pathogen, infecting hundreds of different amphibian species and driving some of them to extinction.

Most important, Bd is our first example of a currently *emerging* pathogen. The definition of an emerging pathogen is very broad; it essentially means a pathogen that we are newly concerned about for some reason. The pathogen may be truly novel, emerging through mutation. However, the most common cause of emergence is the transfer of existing pathogens to a new host species (which we call zoonosis when the transfer is into humans). Alternatively, emergence might describe an increase in the virulence or transmission of an existing pathogen (due to mutation, or to some change in the host or the environment). Finally, a pathogen might emerge as host tolerance or resistance declines due to a changing environment.

For emerging pathogens of humans, the prime suspect is zoonosis (as in HIV), sometimes combined with pathogen mutation (as in pandemic influenza). Human pathogens can also emerge when environmental change opens the environmental filter for an existing disease, as when mosquito vectors of dengue spread to North America, or when people increase their contact with Lyme disease-bearing ticks by building houses in wooded areas.

Emerging diseases of non-human species are a concern for several reasons. Many species are economically valuable; emerging disease can threaten our wallets or even our lives. Agricultural disease can threaten important crops. The Irish potato blight—a fungal pathogen that jumped from its origin in South America to Europe via North America—caused massive human mortality and fundamentally altered the history of Ireland. The recent decline in honeybees, the causes of which are still hotly debated but which certainly include pathogens, has severely compromised the California almond crops they pollinate; chronic wasting disease in wild elk threatens to cost Canadian elk farmers millions of dollars.

Diseases can also affect economically important wild populations, especially when pathogens are passed back and forth between wild and farmed populations of the same species. Sea lice, a parasitic crustacean of fish that thrives in high density salmon farms, may be spilling over to harm wild salmon populations. Hopefully, we would also care about the welfare of non-human organisms for selfless reasons—conserving non-human species is simply the right thing to do.

Bd in amphibians is definitely in the latter category—despite the economic value of some amphibian species in the frog-leg trade or in controlling pests, most of the species affected by Bd are economically unimportant. Nevertheless, we want to understand where Bd came from and discover how we can protect wild amphibian populations from its impact. The concepts and strategies we use in our efforts to determine Bd's origin can also serve as a case study for understanding and controlling future emerging diseases of non-human species.

Physiology and natural history

Bd is the black sheep of a large and otherwise obscure family of fungi, the chytrids. While a few chytrid fungi attack other fungi or plants, most live harmlessly on decaying organic matter in aquatic environments. Bd is one of only two known chytrids that attacks vertebrates, the second being a salamander pathogen that affects many fewer species. (Chytrid biologists complain when people refer to Bd as '*the* chytrid fungus', feeling that this unfairly taints all chytrids with the misbehaviour of one species.) Bd lives on and within the skin of amphibians, especially on keratin, a hard protein found in the skin. Its life cycle alternates between structures called *thalli*—bottle-shaped cells that grow within the host's skin layers—and *zoospores*, small mobile cells that disperse from the thalli into the water, eventually landing either back on the same host's skin or on another host, thus spreading the infection.

We know very little about when and how Bd persists in the environment, away from its host organisms. This is a critical question if we want to understand how Bd spreads from one amphibian population to another; whether populations can recover or recolonize years after they have been locally extirpated by Bd; and how to design quarantine programmes to protect healthy populations from Bd. Given that many of Bd's relatives are free-living microbes, it would not be surprising if Bd also retained the capability to survive and grow in the environment. We know that Bd can persist in pure water for weeks or months, may be able to survive in humid environments such as 'cloud forests', and thrives on the keratin found in amphibian skins, birds' feathers, and the exoskeletons of insects and crustaceans such as shrimp and crayfish; it can also survive on the keratin in birds' feet, at least as long as they stay wet—suggesting that waterbirds may act as dispersal vectors.

Bd can persist on tadpoles between seasons. This life stage is more tolerant because tadpoles only have keratin around their mouths, which they can discard in response to infection without dying. This form of *intraspecific reservoir*, where pathogens persist in a tolerant life stage of the same species, has also been suggested for other amphibian pathogens. Many kinds of tadpoles grow more slowly but don't die when infected with Bd. Individuals from tolerant host species, which can harbour the fungi and pass them to new (or recovering) populations without being harmed themselves, are potentially important reservoirs and vectors for Bd. Most amphibian communities contain tolerant hosts—researchers are still trying to understand what makes a host tolerant or intolerant of Bd. Some important potential players are the American bullfrog, and the South African clawed frog *Xenopus laevis*. All of these species are widespread, have been able to invade new geographic regions with or without human help, and are tolerant of at least some strains of Bd.

Most current evidence suggests that infected adult frogs die when the build-up of Bd in their skin and its subsequent thickening and hardening prevents them from maintaining proper salt concentrations in their bodies, leading to death by cardiac arrest. Some experiments have suggested that Bd may also produce toxins that contribute to mortality.

Amphibian species vary wildly in their susceptibility to Bd: wide variation in space, time, across species, and across communities is a hallmark of this emerging disease. Some of this variation may stem from differences in host defence. While amphibians have adaptive immune systems that could in principle recognize and fight off Bd in the same way that our immune systems resist bacteria and viruses, there is only weak evidence at present that these systems protect against the fungus—possibly because Bd can produce chemicals that kill or inhibit frogs' T cells (the same cells that HIV infects in humans). The evidence is stronger for the protective effect of the antimicrobial proteins that many species of frogs and toads secrete on to their skins; species whose skin secretions inhibit Bd in a test tube also tend to survive infection better.

Finally, some host species can generate behavioural 'fevers' in response to Bd. Higher temperatures actually increase Bd growth rates in cell culture, but they also enhance frog resistance and tolerance. Although frogs are cold-blooded and can't shiver to raise their body temperatures, infected frogs can and do boost their body temperatures by spending more time in warm, sunny places, which appears to help them survive Bd infections. Frogs can even be cured of Bd in the lab by putting them in warm environments for less than a day. Not all species can be cured in this way, suggesting that rather than harming Bd directly, warm temperatures may help frogs by improving their ability to produce antimicrobial proteins.

Ecologists discovered Bd in the late 1990s when frogs throughout eastern Australia and Central America started dying from mysterious causes. At about the same time, poison dart frogs in the US National Zoo also started dying. Veterinary researchers there got in touch with the few researchers in the world who knew anything about this previously obscure family of fungi. Between them they came up with a species description and a name based on the Latin name of poison dart frogs, *Dendrobates*: the genus name *Batrachochytrium* means 'chytrid that infects frogs and toads'.

One interesting sidelight of the discovery of Bd is the experience of Joyce Longcore, a chytrid expert. After keeping house and raising children for 20 years, Longcore went back to graduate school, receiving a PhD in mycology (the study of fungi) in 1991. As of 1997, just before the discovery of Bd, she looked set for a quiet career studying an obscure family of fungi. When Bd suddenly exploded in importance, she was the go-to person for information about the biology of chytrid fungi, and her career took off like a rocket. Since 1998 she has co-authored 87 papers with more than 7,000 total citations, making her a scientific star. Her story demonstrates the value of having the right knowledge at the right time. It also shows that seemingly arcane biological knowledge can suddenly become vital to understanding a novel ecological situation.

Once it became clear that Bd was a previously unknown pathogen, biologists asked about its origins: had it arrived recently in the communities it was destroying, or had it lain dormant in those communities for millennia before suddenly beginning to cause harm? The ensuing debate between the *novel pathogen hypothesis* (NPH) and the *endemic pathogen hypothesis* (EPH), versions of which apply to most emerging diseases of wildlife, has been raging ever since. The argument has recently been tentatively resolved in favour of the NPH, after several decades of ecological and genomic research during which scientists analysed the complex genome of Bd (about the same size as malaria's 23 million base pairs); expanded the range of ecological sampling in space by finding Bd in previously unexplored regions; and expanded the sampling range in time, by retrieving Bd from frog specimens stored in museums for more than a century. As always in biology, there are interesting twists and turns, which we will explore in the rest of this chapter.

The NPH does not say that Bd is a new species—we know that it existed long before the 1990s. Rather, it says that Bd, or at least virulent strains of Bd, is new to the specific geographic areas in which populations are now collapsing due to Bd infection. If the NPH is true, then Bd must have moved into new areas around the time when disease-related die-offs were first observed. Under the NPH we might expect to see a clear spatial separation between regions where Bd has and has not yet arrived, rather than a patchwork of local regions with and without Bd-related die-offs. We should also be able to see a signature of its rapid spread in the spatial pattern of its genomes, with high genetic diversity in regions where it has persisted for a long time and low genetic diversity in areas with recent disease outbreaks. Outside Bd's ancestral home the spatial distribution of Bd genes should be patchy, reflecting the haphazard processes of dispersal.

In contrast, the EPH asserts that the same strains of Bd have been present in amphibian communities, even the ones now experiencing disease outbreaks, for a long time. The *disease triangle*, an idea from plant epidemiology, says that a disease outbreak requires the presence of (1) a suitable host, (2) a pathogen, and (3) an environment in which the pathogen can overcome the encounter and compatibility filters in order to successfully spread from one host to another, and overcome resistance and tolerance to cause disease. The EPH says that the first two sides of the triangle have been in place for centuries or millennia, but that some change in the environment has recently opened the encounter and compatibility filters, or changed tolerance. Since we believe Bd must be able to infect amphibian hosts in order to persist, it is not the encounter or compatibility filter that has opened; rather, proponents of the EPH think that changes in the environment have made Bd more virulent or hosts less tolerant of Bd. To validate the EPH, we should not only be able to reject the predictions of the NPH by showing that the spatial pattern of genetic variation in Bd is geographically structured; we should also be able to identify environmental covariates that predict Bd-drive die-offs, and these covariates should have changed recently in outbreak regions.

Before we go through the evidence for each hypothesis, it's worth keeping in mind that Bd emergence is complex—like all biological phenomena—and that the NPH and EPH are not mutually exclusive. It could be true both that Bd has recently moved into new geographic regions (as stated by the NPH), *and* that Bd has become more virulent or hosts have become less tolerant (as stated by the EPH).

In the early days of the Bd pandemic—just after the discovery of the declines in Central America and Australia, as it became obvious that the impact of Bd varied drastically from one community to another—researchers scrambled to identify environmental changes that would support the EPH, explaining the (apparently) sudden but highly spatially variable virulence of Bd. Puzzlingly, Bd-induced die-offs often occurred in high elevation, pristine areas such as nature reserves—inconsistent with a story involving human-induced changes to the landscape. Two environmental factors that researchers initially thought might be interacting with Bd to cause die-offs, by stressing amphibians or depressing their immune responses, were pesticide contamination (possibly blown long distances from agricultural areas) and ultraviolet radiation (consistent with effects of elevation). Despite some associations with population declines, and some lab studies that have shown that ultraviolet radiation and pesticides can make Bd more virulent, these factors have not (yet) provided much power for predicting where and when Bd will strike amphibian communities.

Temperature has a stronger signal: it has been consistently associated with harmful effects of Bd both on individual animals in lab studies and on communities in nature. As already mentioned, Bd is most harmful to amphibians within a narrow thermal window, especially in the cooler temperatures associated with high elevation tropical forests. While this may explain why lowland communities persist while their uphill neighbours are destroyed by disease, it doesn't explain the temporal patterns—why have high elevation communities only begun to be destroyed in the last few decades and not before, if Bd has been present for centuries? What has changed?

The most obvious candidate, and one that is always on environmentalists' radar, is human-induced climate change. Perhaps climate change, which we know severely affects ecological communities at high latitudes and high altitudes, has recently shifted conditions past a tipping point that allows Bd to spread and/or harm communities, either by increasing the growth rate of Bd (e.g. the zoospore production rate) or by lowering the tolerance or resistance of individual amphibians.

The climate change hypothesis has been controversial within the Bd research community. A high profile 2006 study suggested that reduced daily temperature variation in the Central American highlands had facilitated outbreaks by allowing Bd to remain in its optimal thermal range more of the time. Other researchers claimed that this study made mistakes about important details such as the lag between temperature changes, Bd outbreaks, and community die-offs; a later re-analysis of the data showed that 'numerous other variables, including regional banana and beer production, were better predictors of...extinctions'. Furthermore, the spatio-temporal pattern of the die-offs—which spread across Central America at a rate of hundreds of kilometres per year rather than affecting the entire region simultaneously—seemed more characteristic of the spread of a novel pathogen than of a change in regional climate. Subsequent analyses suggested that changes in temperature variability, rather than mean temperature, were indeed associated with the die-offs, even after taking the spatial pattern of spread into account, and that die-offs might have emerged due to the combination of Bd and the local effects of the global decade-scale climate shift called El Niño, rather than from longer-term human-induced climate change. A global study of Bd die-offs found that annual precipitation is correlated with die-offs, although other similar studies failed to find a strong effect. While changes in climate may well contribute to the occurrence and impact of Bd outbreaks, they do not provide the smoking gun that the proponents of the EPH are looking for.

While the EPH depends on environmental correlations, evidence for the NPH is based on historical and genetic evidence, which should help us determine when Bd arrived in different locations around the globe. As with nearly all infectious diseases of wildlife, Bd existed long before biologists noticed it in amphibian communities. Unlike human diseases such as malaria or influenza, however, there are no historical records that can tell us where Bd was in the past—even if there had been ancient plagues where frogs and toads started dying in huge numbers, humans might not have noticed or recorded it in their journals. In fact, even when people *did* notice amphibian die-offs in the recent past, such as the disappearance of boreal toads from the mountains of Colorado (USA) in the 1970s, or the decline of *Atelopus* frogs in Central America in the late 1980s—declines that we attribute retrospectively to Bd—they attributed them to other causes, such as climate change or environmental stress coupled with bacterial outbreaks.

With the advent of good methods for extracting and amplifying ancient DNA—similar to those used in the archaeological discoveries of malaria in Egyptian mummies—researchers can go back in time by finding Bd DNA on the skins of frogs borrowed from the collections of natural history museums. In both Colorado and Central America, they successfully recovered chytrid fungi from frog skins collected around the time of the outbreaks, making Bd-induced die-offs extremely plausible in hindsight. In the case of Central America, researchers have been able to use earlier specimens to show the *absence* of Bd before the outbreaks, supporting the NPH.

The discovery of Bd DNA on the skins of a Bd-tolerant toad species (the African clawed toad, *Xenopus laevis*) collected in South Africa in 1938 led researchers to propose the 'Out of Africa' hypothesis—the idea that the strains of Bd that would later originate the global Bd pandemic evolved in Africa before 1938 and only began to spread globally after a human pregnancy test based on injecting female *Xenopus* with women's urine was developed in the 1930s, leading to the export of thousands of *Xenopus* a year from South Africa. According to this hypothesis, the fungus would have had two decades to spread globally during the 1940s and 1950s (before other pregnancy tests replaced the *Xenopus* test), after which it gradually spread in new habitats, perhaps via other tolerant hosts.

However, discoveries of the earliest Bd DNA are fragile—every time someone sets a new record by discovering Bd on an even older preserved amphibian skin, the story changes. Since the Out of Africa hypothesis was first proposed, researchers have detected Bd on the skin of an American bullfrog collected in 1928, in California, and from even earlier specimens from Mexico and Brazil going all the way back to 1894.

While researchers will keep pushing back these records, finding earlier and earlier occurrences of Bd around the globe, new discoveries get progressively harder as we go back in time simply because we have fewer museum specimens to test. In the end, more comprehensive geographic sampling of current-day Bd, combined with thorough genomic analysis, has resolved the NPH vs EPH debate in favour of the NPH. While early samples from the Americas, Africa, and Australia failed to determine Bd's origin, continued global sampling and genome sequencing finally localized Bd's original home in East Asia 10,000–40,000 years ago, in host populations that generally tolerated the fungus well. From this ancient gene pool, a *global pandemic lineage* emerged from Asia between 50 and 120 years ago, spreading worldwide probably by means of commercially traded, tolerant species like American bullfrogs and *Xenopus*. This newly virulent lineage (the 'novel pathogen' of the NPH) drove the Bd-driven amphibian declines in Colorado in the 1970s, Central America in the 1980s, and eastern Australia and Central America in the 2000s.

While improved sampling of genomic data may have resolved the NPH/EPH debate, there are always more puzzles to solve—especially the reasons why some host populations are destroyed by Bd while others thrive. Genomic data from amphibian hosts can provide evidence about past population bottlenecks; looking at the genes for antimicrobial proteins can suggest which species needed to defend themselves from Bd in the past. On the EPH side, proxies for past climate such as the isotopic composition of trees or soil (which reveal patterns of temperature and dryness) can give us a more complete picture of the way the host–pathogen–environment disease triangle worked in ancient environments.

Does resolving the NPH vs the EPH help us control Bd outbreaks or lessen their impact on amphibian communities? Unfortunately, not very much, for two reasons. First is the variety of outcomes when amphibian populations encounter the global pandemic lineage of Bd: although the NPH broadly explains the emergence of Bd, the environment is still important at a local scale. Second is our extremely poor ability to control natural systems. *If* we can understand exactly which aspects of the environment control susceptibility of a particular host species, and *if* we can intervene to remove Bd or modify the environment, then we can save amphibian populations. But it's extremely difficult to remove Bd from the environment once it has arrived, and we can't change the occurrence of El Niño events or anthropogenic global change (at least not on a useful time scale). Conservation actions must be based on the overlap of actions that we think will attack the root causes of disease emergence, actions that are ethical (is it OK to cull some members of an endangered species to save the rest?), and actions that are logistically feasible.

We can close the encounter filter by transporting individuals from the wild into disease-free, artificial habitats. The Amphibian Ark is a project that aims to preserve species by making sure that, no matter what happens in nature, we have some animals in captive breeding programmes that can (hopefully) be reintroduced into the wild once we have figured out how to control Bd. Biologists are already successfully rearing endangered species in captivity, but we don't know when or how we'll be able to reintroduce them. Captive rearing buys time, but eventually we either need to breed Bd-resistant or Bd-tolerant variants of these species, or find a way to control Bd in nature. It also raises ethical issues: might we be further harming endangered species by removing individuals from the wild? Could the existence of captive populations reduce the urgency we feel to deal with the problem? Is it OK to raise funds for conservation by raising additional captive animals for sale?

The next most direct way to try to save amphibians is by closing the compatibility filter for individuals in the wild. Antifungal drugs and temperature-raising treatments have worked to cure frogs of Bd in laboratory trials, and we may be able to scale these treatments up to capture, treat, and release enough individuals to save wild populations. Biologists have also proposed to treat amphibians with skin peptides from resistant species, or to treat individuals or even entire communities with a bacterium that produces antifungal compounds that suppress Bd. Researchers have discovered that exposing some amphibians to dead Bd in the lab can give them partial resistance to Bd, but it is unclear whether this finding can be used to develop a practical disease-fighting strategy.

Amphibian species that inhabit single islands are especially vulnerable to extinction (because their populations are often small, and cannot be recolonized if the local population dies out), but conversely they are among the few examples where intensive conservation efforts have made a big difference. Researchers successfully eradicated Bd from populations of endangered Mallorcan midwife toads by removing the toads from their ponds, treating them with chemicals, and simultaneously dosing the ponds with a broad-spectrum disinfectant; the toads remained Bd free after they were returned to the wild. 'Mountain chickens', a Caribbean species of frog, were probably eradicated from the island of Montserrat; they are carefully being reintroduced to the island, although (as is typical of threatened species) biologists are trying to manage a broad range of threats—volcanic activity, overhunting, invasive species, pollution, and development—in addition to Bd. Since endangered species are rarely driven extinct by a single threat, these strategies can help them survive (and eventually evolve resistance or tolerance on their own) even if Bd persists in the population.

The fight against Bd faces the same two fundamental problems as every other disease control and prevention programme—lack of knowledge and lack of resources. No matter how cute or interesting the victims are, diseases of wild animals will never command the same interest as human disease, so we will always have fewer resources—and less knowledge, since resources are needed to acquire knowledge. We do have some advantages—we can cull animals if it looks like it will help us to control an outbreak, breed animals for disease tolerance or resistance, and induce experimental infections to evaluate treatments (all strategies that would be looked on unfavourably for human pathogen control). We haven't yet had to worry about evolutionary countermeasures taken by the pathogen, although these are bound to happen once we start to take action.

Although infectious diseases of wild animals differ from those of humans in superficial ways, the physiological, ecological, and evolutionary dynamics that drive them are very similar. In the long run, analysing wildlife disease helps us understand the fundamental properties of infectious disease. It can help protect harvested or hunted populations that are economically valuable. Understanding how disease moves in natural populations may also provide early-warning systems to detect zoonotic diseases that can jump into humans. Perhaps most fundamentally, many biologists (including the authors of this book) feel that we have an ethical responsibility to preserve species when we can, especially when our actions may have contributed to the spread of diseases that threaten them.

Chapter 8
SARS-CoV-2/COVID-19

As every contemporary reader of this book will know, the biggest infectious-disease story of the early 21st century is the COVID-19 pandemic. Other infectious diseases have hurt or killed more people in total: current estimates of the total worldwide mortality due to COVID-19 (up to the end of 2021) are in the range of 20 million people, while the HIV pandemic has killed 36 million people; the 1918–20 flu pandemic killed 20 to 50 million people in a world with only a quarter the size of today's human population. COVID-19's global impact comes from its transmissibility, its sneaky mode of transmission, and its intermediate virulence (bad enough to be a serious problem, not bad enough to frighten people into compliance with public health measures). In contrast to most of the other modern-day plagues described in previous chapters, it is not 'somebody else's problem'; it affects sexually monogamous people in high as well as low income countries who do not use intravenous drugs, who can afford clean water, and who live in temperate climates outside the range of most disease vectors. Although we can understand many features of COVID-19 epidemiology using the same general principles we have discussed in previous chapters, this pandemic has underscored the importance of human behaviour.

The virus that causes COVID-19 is called SARS-CoV-2. 'SARS' stands for 'severe acute respiratory syndrome'—referring to

the symptoms of the disease, similar to those caused by the
SARS-CoV-1 virus that emerged from China in 2003, infecting
8,000 people and killing 774 worldwide. 'CoV' stands for
'coronavirus', a large family of RNA viruses. Like influenza, the
virus directly uses the biochemical machinery of its host cell to
create new virus particles. Prior to the emergence of SARS-CoV-1,
coronaviruses were known as viruses of animals such as mice and
chickens, and as an occasional cause of seasonal respiratory
infections (i.e. the 'common cold') in humans. SARS-CoV-1 was
successfully contained and disappeared from the human
population within a year. The period between the SARS epidemic
in 2003 and the beginning of the COVID-19 pandemic in 2019
also saw the emergence of another human coronavirus, Middle
Eastern Respiratory Syndrome (MERS), which spills over into
humans from camels but spreads poorly from human to human.

Given the strong similarities of SARS-CoV-1, MERS, and
SARS-CoV-2, why did only SARS-CoV-2 successfully emerge as a
global pandemic? All three pathogens are from the same family,
descend evolutionarily from ancestors found in bats, and most
likely made their way into humans via intermediate hosts—palm
civets, dromedaries and camels, and (possibly) pangolins
respectively—although we are still uncertain about the
intermediate hosts for SARS-CoV-1 and 2. And all three can cause
severe illness and death in humans.

The clearest difference among the three pathogens is in their
virulence, as measured by the *case fatality rate* (CFR), the
probability that someone with a reported case of the disease
will eventually die from it: the CFRs for emerging human
coronaviruses range from about 2 per cent for COVID-19
(in unvaccinated people), to 15 per cent for SARS, to 35 per cent
for MERS. All three of these numbers represent extremely rough
approximations, with risks varying by as much as 10-fold by age
(with older people at higher risk in all cases) and across space and
time. CFR depends heavily on healthcare capacity, both for the

obvious reason that better care can save an infected person from dying and for the less obvious reason that better infectious disease surveillance will detect a larger number of people with mild or asymptomatic infections. Since the CFR is the ratio of mortality to the number of cases reported, higher reporting of mild cases will reduce the CFR.

The CFR takes account of only reported cases. If we want to know how dangerous a disease really is, we may need to estimate the *infection fatality rate*, which is the probability of death for someone who gets infected, whether or not their case is reported to public health authorities. Estimating the infection fatality rate is tricky, because we cannot use public health statistics alone. Two common approaches are to estimate how many people were infected by taking blood samples and testing them for the presence of antibodies against the virus, or finding a setting where we think that all infections would have been detected. For COVID-19, one such famous example was the February 2022 outbreak on the *Diamond Princess* cruise ship, where many of the passengers underwent random testing whether they had symptoms or not. (However, we need to be careful extrapolating this infection fatality rate to the general public since people who go on cruises are generally older, and therefore more likely to die from COVID-19, than the general public.)

A further source of uncertainty is that mortality due to disease is often underreported. While it is unlikely that any COVID-19 deaths on the *Diamond Princess* were missed, deaths due to disease in the general population are often misattributed, especially when someone dies of secondary causes such as bacterial pneumonia several weeks after their primary infection ends. Epidemiologists measure *excess mortality* by comparing the reported number of deaths from all causes to the average mortality rate in that week of the year. The discrepancy between excess mortality and reported disease deaths can be especially high in regions with poor reporting systems; for example, Zambia reported fewer than 4,000 COVID-19 deaths in the entire pandemic (to June 2022), but excess mortality for 2021 alone represented 80,000 deaths.

Among the emerging coronaviruses, we see that the more virulent diseases such as MERS are also the least transmissible. The R_0 value for MERS is close to 1 (for transmission between humans) and for SARS is around 3. COVID-19's R_0 was initially around 3 but increased to around 8 with the emergence of later, much more transmissible strains. While much of the variation in R_0 between diseases comes from complex virological and immunological differences, one major factor that helps explain both the transmissibility (R_0) and virulence (infection fatality rate) of coronaviruses is whether a particular species or strain replicates better deep in the lungs or in the upper respiratory tract. Viruses such as MERS that favour the lower respiratory tract are more likely to compromise lung function and lead to severe disease and death, but their infectious particles are less likely to escape from the host's airways into the environment where they can infect others. This tradeoff between virulence and transmission partly explains the trends in transmission and virulence between MERS (most virulent/ least transmissible), SARS-CoV-1, and SARS-CoV-2 (least virulent/ most transmissible). It may also explain the high transmission and low virulence of the late evolving Omicron strain of SARS-CoV-2 relative to earlier strains. However, we should not take this particular difference as proof that pathogens in general, and SARS-CoV-2 in particular, will always evolve towards lower virulence; there are so many possible biochemical innovations, and so many ways to interact with the human immune system, that nasty surprises are always possible. For example, the second (autumn) wave of the 1918 influenza pandemic was considerably more virulent than the first (spring) wave; biologists suspect that this was due to mutation rather than changes in the environment, but we may never know.

While SARS-CoV-1 and the initially emerging strains of SARS-CoV-2 had similar R_0 values, SARS-CoV-2 is much harder to control because it can more easily jump from an infected to a susceptible person before the infected person starts showing symptoms (presymptomatic transmission). Indeed, it can even be transmitted during a relatively mild infection where the infected person never shows symptoms at all (asymptomatic transmission). This difference isn't reflected in the viruses' R_0 value, because R_0 measures the transmission potential in a population without any attempts by individuals or governments to lower transmission. Early in a pandemic people may not pay careful attention to cold- or flu-like symptoms, and as a result can easily transmit their infection; as the pandemic goes on and people are alerted to the dangers of infection, any pathogen that relies on transmission after the onset of symptoms will find itself in a dead end as infected people are quarantined, either voluntarily or involuntarily. The low virulence of SARS-CoV-2 (relative to SARS-CoV-1 and MERS, not to other diseases like influenza or measles!) may also have made it more difficult to control, as it is harder to convince people to undergo severe inconvenience to stop the spread of a disease that to many feels like 'just a cold'.

Although we know that SARS-CoV-2 spilled over into human populations sometime in late 2019 (hence the name COVID-19, although the pandemic did not spread widely until 2020), scientists are still (as of mid-2022) bitterly divided over whether it came from bats to humans through wildlife being sold in a live market (the 'natural origin' hypothesis), or whether it was accidentally released from a virology lab (the 'lab leak' hypothesis). There are three versions of the lab leak hypothesis. The first, that SARS-CoV-2 was deliberately engineered as a bioweapon, has little support except among conspiracy theorists. The second and third versions both say that SARS-CoV-2 was collected from bats in the wild and held in a virological lab (most likely the Wuhan Institute of Virology) before accidentally being released; the difference between them is whether biologists in the lab used gain-of-function experiments (such as those discussed in Chapter 3) to improve its ability to infect humans. Chinese public health agencies were slow to release important information at the beginning of the pandemic—they did not admit publicly that the new virus was human-transmissible until late January 2020, after many infected people had already travelled to other places for the lunar New Year holiday. We will probably never know, however, whether these agencies withheld or covered up other important information. Because natural evolutionary processes and gain-of-function experiments can produce the same changes in a virus, and because we can almost never reconstruct the chains of infection that started an epidemic, the lab leak hypothesis is *unfalsifiable*—we can argue about the balance of probabilities for different sources of the epidemic, but never absolutely rule out the possibility of a lab leak. In a way, this uncertainty may be good; regardless of how the COVID-19 pandemic started, it's clear that either lab accidents or natural spillover could create the next pandemic. The simple conclusion is that we should work *both* to increase laboratory biosecurity (including thoughtful conversation about the risks and benefits of gain-of-function experiments) and to minimize opportunities for zoonotic spillover (and in particular, to prevent spillovers from developing into human pandemics).

However it made its way into the human population, analysing SARS-CoV-2's genome tells us that it emerged from bats (possibly by way of another mammal such as pangolins), and that we haven't yet found its close relatives in the wild. The nearest relative we have found, the bat virus RaTG13, apparently diverged from SARS-CoV-2's ancestors about 50 years ago. Detailed analysis of the genomes from early COVID-19 cases, accounting for the fact that recombination in coronaviruses can occur anywhere in the genome (not just by reshuffling of discrete genomic segments, as in influenza), suggests that the COVID-19 pandemic actually started from two separate zoonotic events in late November or early December of 2019. It furthermore suggests that successful spillover events—those resulting in epidemics or pandemics among humans—are just the tip of an iceberg of zoonoses, with most chains of infection dying out before any public health agencies notice them. However, further sampling may eventually change our conclusions about SARS-CoV-2's origins, just as it did for malaria and Bd.

SARS-CoV-2 continues to evolve, having entered the human population. Since SARS-CoV-2 has only recently colonized humans, there is still plenty of scope for mutations that improve its adaptation to humans, particularly via increased transmissibility; the variants emerging in the first year of the pandemic (Alpha, Beta, and Gamma) all showed mild to moderate (50 per cent–200 per cent) increases in transmissibility relative to the initial strains. We should expect to see this pattern in any emerging disease, as the pathogen searches for its evolutionary peak. While we did not observe such clear increases in transmissibility early in the HIV pandemic, much of HIV's adaptation to humans probably occurred in the decades before it was discovered.

On the other hand, the early emerging variants of SARS-CoV-2 did not display the negative correlation between transmission and virulence expected from comparisons of SARS-CoV-1, MERS, and SARS-CoV-2. For example, the Delta strain appears to have been both more virulent and more transmissible than the strains before it. Increased transmission always helps the parasite, but increased virulence will only be harmful if it cuts short the pathogen's infectious period: by killing the host, provoking a more effective immune response, or changing host behaviour to reduce transmission. Only with the emergence of the Omicron strain, which is both much more transmissible than earlier strains and less virulent, did we see evidence of the same pattern of upper-respiratory mild virulence and high transmissibility observed more broadly across human coronaviruses.

Omicron's other important evolutionary innovation was its ability to hide—at least partially—from immunity previously acquired either by vaccination or natural infection. While earlier strains emerged before vaccines were available, and before a large fraction of the population had been infected by SARS-CoV-2, Omicron also occupies a distant evolutionary branch that is far away from previous strains. Scientists are still debating where and how it changed so much: the leading hypotheses are that Omicron evolved either (1) in a population of non-human mammals ('spillback'); (2) in some remote geographic region where SARS-CoV-2 circulated without anyone having a chance to sample and sequence its genome; or (3) in a prolonged infection of one or more immunocompromised humans.

The evolutionary future of SARS-CoV-2 is still a mystery. While virologists and evolutionary biologists can often explain in hindsight why a pathogen evolved in a particular way, forecasting evolutionary changes such as which mutations will occur, and how they will interact with each other and with the human immune system, is almost impossible. Our best guess is that future SARS-CoV-2 will look similar to influenza, with moderate evolutionary costume changes (antigenic drift) occurring every year, and with the unpredictable occurrence of more radical costume changes (antigenic shifts in the case of influenza, jumps like those of Omicron for SARS-CoV-2) that reduce the population's previous immunity. SARS-CoV-2 may be even more infectious from year to year than influenza, as our immune systems have a harder time 'remembering' coronaviruses they have seen before. Predicting the evolution of virulence is even harder: perhaps selection for increased transmission will continue to push SARS-CoV-2 to replicate higher in the respiratory tract, decreasing its virulence—or perhaps other mutations will increase its virulence (despite decades of research, we still don't understand why the 1918 pandemic influenza virus was so terribly virulent).

Since SARS-CoV-2 will be with us for the foreseeable future, we need to understand how to control it and mitigate its effects. Non-pharmaceutical interventions (NPIs)—control measures that rely on closing the encounter rather than the compatibility filter—are the first line of defence when we are faced with a new pathogen, and SARS-CoV-2 was no exception. Because NPIs require people to change their behaviour (wearing a mask, washing hands) or make economic sacrifices (cancel public events, lose pay by staying home when sick), they are hard to sustain over time.

Many public health researchers would say that the COVID-19 pandemic has not changed our previous understanding of infectious disease in any fundamental way, but it has certainly reinforced lessons we should have learned earlier. One example is the effectiveness of quarantine via border controls, either banning travel completely or quarantining arriving travellers. Such measures are politically popular; they inconvenience only a small fraction of the population, few of whom vote in the country that is imposing the controls; and they reinforce the narrative that a pandemic is someone else's problem. But because all but the most draconian controls are leaky, they are most useful for prolonging the period before the pathogen establishes locally, buying time for planning and developing treatments or vaccines. Once the pathogen inevitably sneaks in, the epidemic quickly grows to the point where the risk of infection from someone in the community dwarfs the risk from foreigners. In many cases, and especially with cryptic pathogens like SARS-CoV-2, border controls are imposed only long after they would have been useful.

National or regional lockdowns, at various levels of stringency from cancellation of large indoor public events to city- or country-wide prohibition of all but necessary activities outside the home, can go a long way towards closing the compatibility filter. However, they are unpopular and economically, socially, and psychologically harmful: they exacerbate economic inequities, damage mental health, and interfere with children's education. The effectiveness of lockdowns is highly contentious and extremely difficult to quantify. In order to know how many disease cases a lockdown prevented, we need to know how much people would have voluntarily restricted their activity even without a lockdown; how much they actually complied with the rules of the lockdown; and what other phenomena were simultaneously changing the potential for disease spread (e.g. seasonal variation in transmission due to parasite biology and human behaviour, decreases in susceptibility due to natural infection, or parasite evolution).

To avoid an all-out lockdown, one could try to track the movements of infected people and encourage or require quarantine of people they may have infected (contact tracing). This strategy, wherein epidemiologists try to get ahead of an emerging outbreak and find everyone who is infected before they have a chance to infect too many others, works well for slowly reproducing diseases such as HIV, and those such as Ebola that have unmistakable symptoms associated with infection, but less well for diseases that have short generation times or presymptomatic and asymptomatic transmission. Even in the best-case scenarios, contact tracing breaks down as soon as a disease spreads widely, because the number of cases and the number of people who must be found and observed or isolated snowballs beyond the ability of a limited number of contact-tracing personnel to keep up. Early in the pandemic, epidemiologists were hopeful that digital contact tracing using cellphone applications could overcome these problems. While this strategy is still promising, it depends on voluntary participation (or levels of government control that are unacceptable in democratic societies). In practice, only a small fraction of the population used the apps; their use may have averted hundreds of thousands of cases, but these estimates are highly uncertain.

Milder forms of behaviour change that partially close the encounter filter, such as masking, handwashing, or physically distancing, are less onerous but also less effective. And, when less effective measures become policy, this can undermine public trust. The same questions that plague estimates of the efficacy of other NPIs also obscure our understanding of behavioural mandates: how much do people comply with mandates, and how much would they have worn masks even without the mandate? What else about the pandemic was changing at the same time?

One bright light among all of this darkness and mystery is the recent set of breakthroughs in the vaccine industry. Prior to the COVID-19 pandemic, the shortest time from novel pathogen identification to an effective vaccine was 10 years (for measles). Previous vaccine technologies depended on culturing viruses in the lab and inactivating them with heat or chemicals; evolving milder (attenuated) strains of virus; or inserting genes that produce viral proteins into bacteria to manufacture the proteins in the lab (recombinant vaccines). The development of mRNA vaccines, which use viral messenger RNA to trick host cells into producing viral proteins in the same way the virus itself would, has vastly accelerated the time scale of vaccine development. The first COVID-19 patients were detected in mid-December 2019; Chinese scientists published a draft version of the genome by 11 January 2020; virologists designed a vaccine within 48 hours; and clinical trials of the vaccine began two months later. Nearly all of the time between the emergence of COVID-19 and the availability of vaccines starting in December 2020 was taken by studies of vaccine safety and efficacy required before public health agencies could approve the vaccines for emergency use, rather than vaccine development itself. Development of vaccines against new strains of SARS-CoV-2 will be even faster since they are minor variations on now-tested vaccines, although lead times of at least months will still be needed to prove that the vaccines work, as well as to produce and ship millions of doses to where they are needed.

In theory, vaccination should be more sustainable than NPIs—one need simply visit the doctor for a pill or jab. Every silver lining has a cloud around it; the surprise for public health practitioners was how hard it was to convince people to get vaccinated with a safe, effective vaccine in the middle of a pandemic. Perhaps this reluctance should not have come as such a surprise—even setting aside conspiracy theories and the politicization of vaccines, vaccination has been a hard sell ever since the people of Boston resisted Cotton Mather's call for smallpox inoculation in 1721. More generally, the COVID-19 pandemic has reminded public health agencies that the final frontiers of disease control are not biological: they are political, sociological, and psychological.

Chapter 9
Looking ahead

This book has given a whirlwind tour of some important infectious diseases, their ecological and evolutionary principles, and how these principles inform treatment and control. We chose the few case studies we could fit into the book on the basis of socioeconomic importance and ecological/evolutionary interest, covering a broad range of disease-causing taxa. We chose diseases we thought would be familiar to our readers.

We have regretfully omitted many infectious diseases that hurt many people, costing hundreds of thousands of lives, millions of disability-adjusted life years, and billions of dollars. Some examples include tuberculosis (arguably the most important disease we have neglected), polio, and schistosomiasis (a parasitic disease that causes liver damage, mostly affecting people in sub-Saharan Africa). So many such diseases are rampant in middle and low income, tropical countries that the World Health Organization has defined them as their own category, 'neglected tropical diseases', and there is a scientific journal devoted to them.

We have also left out fascinating diseases that have shaped history. The bubonic plague, the greatest infectious killer ever, is now relatively easy to treat with antibiotics. Smallpox, the first disease to be eradicated in the wild by vaccination, destroyed native populations in the Americas and facilitated European colonization. Rinderpest, a cattle disease closely related to measles, may have transformed the landscape of east Africa by wiping out native wildlife and allowing the growth of shrubby vegetation which allowed tsetse flies, and through them a vector-borne disease called sleeping sickness or trypanosomiasis, to establish. (A successful vaccination campaign led to the global eradication of rinderpest in 2011, the second and so far the last endemic parasite to be driven extinct by humans.)

Only half of our examples (HIV, Bd, and COVID-19) are emerging rather than established diseases; the 2009 pandemic H1N1 strain of influenza could also count as 'recently emerged'. We didn't have room to discuss the henipaviruses such as Nipah virus that threaten to spill over from fruit bats in Australia and South East Asia, or emerging vector-borne diseases such as West Nile and dengue viruses, or the bacteria causing Lyme disease.

Finally, we have covered only a small range of the kinds of organisms that can cause infectious disease. Viruses such as influenza and HIV, bacteria such as *Vibrio cholerae*, protozoans such as the malarial agent *Plasmodium falciparum*, and fungi such as *Batrachochytrium dendrobatidis* do represent the vast majority of pathogenic agents. However, we have passed over multicellular parasites such as roundworms (nematodes) and flatworms (platyhelminths), which have traditionally been thought of separately from microparasitic infectious diseases (see Chapter 2), but which obey the same epidemiological, ecological, and evolutionary principles. Most neglected tropical diseases are caused by protozoans and multicellular parasites. The fact that these diseases mostly affect people in low and middle income countries, as well as their tendency to cause chronic debility rather than acute disease, contributes to their neglect.

The first thing we know is that *plus ça change, plus c'est la même chose* (the more things change, the more they stay the same). Outbreaks of the Ebola virus have already killed tens of thousands of people since its discovery in the 1970s, with the largest epidemics by far occurring in western Africa in 2014–16. The emergence of Ebola underscores many of the points we have made throughout this book. Ebola is a zoonotic virus that likely came to us from bats (like COVID-19), and while the shift from bats to people is uncommon, genetic analysis tells us that it has happened several times. The local postcolonial infrastructure could not deal with the challenges of Ebola alone, and the attention of the world was elsewhere. This neglect changed abruptly when, as with almost every other disease we have discussed, global travel enabled the spread of Ebola to North America and Europe. Thus far, transmission outside of Africa has been stopped promptly. Highly effective vaccines are now available for Ebola, although limited supply has restricted their use to curtailing outbreaks rather than preventing them.

As with HIV, fear of Ebola has stigmatized the disease. People understandably attempt to hide infection in themselves or their families, further complicating attempts to understand the scope and dynamics of the epidemic. Ebola terrifies people in part because of its mode of transmission, and in part because of its extreme fatality rate. All the bodily fluids of a person with an active Ebola infection—particularly anyone who has died from the disease—are loaded with infectious virus. In this way, like cholera, Ebola violates the usual transmission–virulence rules: it is most transmissible when it is most deadly. However, Ebola's grisly mode of transmission means that the encounter filter is narrow. People other than healthcare workers and those who take care of the dead are unlikely to become sick, because it is unlikely that they will encounter contaminated bodily fluids. (Unfortunately, some traditional African customs mean that many of a dead person's relatives come in contact with the body before and during the funeral.) For example, family members of Thomas Duncan, the first person to be diagnosed with Ebola in the USA who subsequently died from the disease, remained healthy despite living with him for several days after he became contagious. Two of the nurses caring for Duncan in a hospital did become infected, but happily recovered.

Zika, another formerly neglected tropical disease, rose to global visibility in 2015 when Brazilian researchers discovered a 20-fold increase in births of children with microcephaly (head circumference much smaller than normal) in regions with high infection rates. Children born with microcephaly may be deaf and blind, have seizures, intellectual disabilities, or other problems. Microcephaly does occur in approximately 1 per 10,000 live births in the absence of Zika infection, caused by other parasites such as toxoplasma or cytomegalovirus, or sometimes for no apparent reason. Even the elevated rate of 20 cases per 10,000 live births may seem small, but so many people overall, in particular so many pregnant women, were infected in the 2015–17 epidemic that Zika caused more than 2,500 confirmed cases of microcephaly—an untreatable, lifelong condition.

Zika is a vector-borne disease, caused by the virus ZIKV and transmitted by the bites of the mosquitoes *Aedes aegypti* and *Ae. albopictus*. These mosquitoes are found all over the world, primarily in tropical and subtropical zones. ZIKV is a positive strand RNA virus, in the same family as dengue and West Nile. It was named for the Zika Forest in Uganda, where it was first isolated from a monkey and described. Most cases of Zika are mild, causing rash, fever, joint pain, or even no symptoms at all. Because early Zika outbreaks were both small and understudied, the more alarming disease outcomes went undetected until the disease had been known for over 70 years. In addition to microcephaly, Zika can cause encephalitis in adults, as well as a rare neurological condition, Guillain-Barré syndrome (GBS). GBS is caused by an immunological reaction to parasites—ZIKV, COVID-19, and malaria among others. Happily, most people make a full recovery from GBS with proper treatment.

How did researchers ultimately make the link between Zika infection in pregnancy and microcephaly? Researchers were perplexed that there had been no reports of increased rates of microcephaly during the previous large epidemic, when 9,000 people were infected in French Polynesia in 2013–14. Had the virus mutated to become more virulent, or was there a new environmental factor such as an insecticide or a vaccine interacting with the virus? The answer was 'neither of the above'. Brazil had a strong enough health infrastructure to detect what had previously been missed: a retrospective analysis revealed that there had indeed been an increase in microcephalic births during the Zika outbreak in French Polynesia.

In the past few decades, researchers have also discovered several new modes of infectious disease that seem almost like science fiction. The first, *prions* or transmissible infectious proteins, are misfolded proteins that can replicate within a host by catalysing the misfolding of other proteins to the prion form. Prion diseases such as scrapie, known to infect sheep since the 1700s, and chronic wasting disease, which was first detected infecting deer in Colorado in the 1960s, are most often transmitted from one animal to another when animals eat vegetation contaminated with prion proteins. Prion proteins get onto plants, completing the transmission cycle, through environmental contamination from animals' bodily fluids (saliva, faeces, or amniotic fluids) or when released into the soil from their cadavers.

Prion diseases hit the headlines in the 1990s with 'mad cow disease', officially called bovine spongiform encephalopathy (the condition is drily called 'variant Creutzfeldt–Jakob disease' when it occurs in humans). Along with the fear of contracting a disease that leads to fatal neurological degeneration, the British public was also fascinated by the grotesque cause of the outbreak, which was due to involuntary cannibalism among cattle. To promote their growth, the animals were fed protein supplements that included brain and spinal cord tissue from other cattle. Prion diseases can extremely rarely occur spontaneously in animals, or due to rare genetic defects; when the remains of these animals are mixed into the food of hundreds of other animals, catastrophe results. A similar but even more macabre epidemic of prion disease, involving human rather than bovine cannibalism, spread through the Fore people of Papua New Guinea starting in the early 20th century, in the wake of their adoption of 'mortuary cannibalism'—the practice of ceremonially eating their dead relatives—and vanished again after the abandonment of cannibalism in the mid-20th century.

What is the outlook for the control of infectious diseases over the next few decades? What are the likely impacts of infectious disease on your health and welfare, or on your family's, from diseases that are already present, or from newly emerging ones? How will the ecology and evolution of disease change in the future?

Plus ça change, plus c'est la même chose. Our understanding of how diseases are transmitted, and the development of vaccines and treatments that can close the compatibility filter, has revolutionized disease management, but the basic processes driving the disease ecosystem remain the same. Humans have had some resounding successes: we have completely eradicated smallpox and rinderpest, and we can realistically consider the possibility of polio and measles eradication, even though the last steps are proving to be immensely difficult for social, political, and economic reasons. Though it remains hard to imagine eradicating HIV, we have developed treatments that allow infected people (at least those with access to good medical care) to tolerate the disease and live out their regular lifespan.

However, we have also lost ground. Tuberculosis has re-emerged, especially in conjunction with HIV; the first optimistic decades of malaria control ended in retreat; and new diseases such as Lyme disease, West Nile virus, H1N1 influenza, and COVID-19 have continued to spill over from animal populations. Perhaps the scariest failure has been the emergence of drug and antibiotic resistance, including methicillin-resistant *Staphylococcus aureus* (MRSA) or malaria or multiply resistant tuberculosis. We are once again forced to contemplate the spectre of untreatable diseases.

A future free of infectious disease is simply unrealistic. Living things have parasitized one another since the beginning of life itself, and no amount of intervention will alter that. New diseases will be created by mutation or recombination of existing ones and by spillover from animal populations, and existing diseases will continually evolve to escape our methods of control. What is attainable, however, is minimizing the impact of disease while understanding that it will always be with us. We *can* slow or stop pandemics, and we *can* reduce the amount of death and misery that diseases cause, even if we can never fully conquer them.

What has changed in our understanding of infectious diseases over the last 50 years? During that period, we moved from population-level treatment to individual treatment. We have learned that population-level treatment, in contrast to curing individuals after they have already been infected, is still invaluable in stopping disease. The simplest way to minimize the impact of disease is to minimize its incidence in the first place. We are learning that there is synergy between population-level and individual-level approaches to prevent disease. The phenomenon of herd immunity is one such example (see Chapter 2).

Our case studies have illustrated that disease-causing organisms are continuously evolving. Tuberculosis is re-emerging in part because the bacteria that cause it have evolved antibiotic resistance. HIV evolved to survive better in humans following its host shift from other primates. But remember—mutations happen at random. So too do encounters between relatively harmless bacteria and bacteria or viruses carrying dangerous pathogenicity genes. Because mutations happen in every organism, the fewer organisms that are present, the less likely it is that they will hit on just the right mutation to cause trouble. Think of it this way: if there are 20 viruses, and a mutation that makes them resistant to an antiviral agent happens only one in a million times, it's improbable that any of them will acquire that rare mutation. But if there are a billion of them, it is practically inevitable. So by keeping the numbers of organisms that infect us low through population-level interventions like effective vaccination or quarantine, we reduce the probability that they will evolve to become more dangerous. Again, an ounce of prevention is worth a pound of cure.

One of the lessons from past failures of disease control is that the miracles of modern molecular biology have limitations. Vaccines are simplest to develop when the human body already has a quick and effective immune response. For diseases like HIV, malaria, or tuberculosis that use evolutionary costume switching, immunosuppression, or other tricks to evade the immune response, we may never be able to achieve the same cheap, effective vaccines that eradicated smallpox and rinderpest, and have brought measles and polio to the brink of eradication. More generally, any control strategy that focuses on a single method, such as trying to eradicate malaria solely by spraying insecticides in the environment, is doomed to failure.

Even as we have learned the limitations of magic bullets, we have developed new technological applications that can help both to detect parasites and to control them. Identifying viruses or testing bacteria for antibiotic resistance used to take days or weeks. Virus classification required careful microscopic inspection or time-consuming immunological techniques. Testing for resistance required assessing the ability of bacteria to grow on plates containing various antibiotics. Now we can diagnose and characterize bacteria and viruses within hours by sequencing their genomes. Moreover, these techniques are no longer relegated to skilled technicians in laboratories; new sequencing technologies can test unprocessed samples (for example, of saliva or blood) under field conditions, and costs are dropping by the month. Such genotyping technologies are also proving to be useful in bacterial infections, for tailoring the correct sort of phage to use to attack them.

We have also learned the power, as well as the limitation, of changing human behaviour. We can easily close the encounter filter for many diseases by checking thoroughly for ticks after we go into the woods, stopping the exchange of bodily fluids with strangers, and staying home (or keeping our kids home) when we develop a cough. But the practical and economic costs of changing our behaviour often mean that we keep exposing ourselves, and others, even when the solutions seem (on the surface) to be relatively simple and even when the consequences (such as contracting HIV) are dire. For you, the reader of this book, this is relatively good news; there are lots of simple ways that you can prevent your own infection, or, if you are infected, stop transmitting disease to others. But humans as a whole are stubborn creatures, and we have many conflicting priorities—earning a living or even saving a little bit of time may often trump the practices that could save us, or others, from infection.

One of the important frontiers in disease control is figuring out better ways to honestly and effectively inform the public about the dangers of disease and the methods that are in their hands for controlling it. Sometimes the interests of the individual and the population diverge. For example, staying home from work when we're sick might be a bad individual decision, because it costs us a day's pay or a boss's goodwill, even if it benefits our co-workers. We must design better policies that encourage compliance with public health goals, such as paid sick leave, using a sensible combination of carrots and sticks without unduly restricting individual freedoms.

The continuing pressure of humanity on the environment is another thing that has changed, but has also stayed the same.

Urbanization and increased population growth increase the rates at which humans contract new pathogens, largely carried by animals. As we move into previously uncolonized habitats, or as we modify our environments, we increase the rates at which we encounter other animals and the parasites they carry. This problem applies to temperate diseases (such as Lyme disease) as well as tropical diseases. Humans are changing their environment in huge numbers of ways—cleaning our water, aggregating in modern mega-cities, clearing tropical forest—that will change our epidemiological landscapes both for better and for worse.

Epidemiologists are intensely debating the possible effects of the largest-scale uncontrolled experiment in history—the release of CO_2 from fossil fuels into the atmosphere, and the concomitant changes in global climate—on disease prevalence. On the one hand, there is no question that ecological change brings about epidemiological change; many species' ranges have already shifted in response to climate change, and insect vectors such as mosquitoes will almost certainly shift as well. In particular locations, such as the highlands of east Africa, increasing temperatures do indeed seem to have driven increases in the incidence of malaria. But the regional effects of climate change are complex, involving changes in variability, seasonal patterns, and the hydrological cycle as well as the overall temperature. Increased human migration and misery (malnutrition, etc.) are probable outcomes of climate change that are also likely to change disease susceptibility and transmission. Add that to the complexity and unpredictability of interactions of ecological systems with regional climate, and scientists note that although we can say with certainty that climate change will have *some* effect on disease, it is hard to know exactly what it will be.

As mosquitoes shift their ranges to affect people in temperate countries, people in those countries will become much more interested in eradicating them. At first glance, getting rid of mosquitoes, at least the ones that we know carry particularly devastating diseases, seems like a good idea. Indeed, it has recently been argued that scientists have become too narrowly focused on problems of drug resistance of malaria, and that we would do better to launch an attack on mosquitoes. Scientists are developing new techniques for mosquito control, like introducing mosquitoes infected with bacteria (Wolbachia) that decrease mosquitoes' ability to transmit disease or make males effectively sterile; these new techniques can supplement or replace clumsy, imprecise interventions like draining wetlands and/or use of pesticides that devastate many other species. These techniques are also more 'evolution-proof' than insecticides, although even they too will eventually be overcome by mosquitoes' evolutionary countermeasures. However, there's a larger point to be made here: we mustn't allow our thinking about infectious disease to stagnate. Overconfidence in any one solution (drug treatment or mosquito control), or overfocus on any part of the multifaceted problem that is infectious disease (e.g. solely on parasite resistance), limits our ability to solve any kind of problem.

Human health is not just about vanquishing wicked parasites, either with magic bullets, everyday practice, or even both. The parasites are embedded in the same natural system that we are. Though we have spent little time discussing it in this book, human sociology, including the legacy of colonialism, also has huge effects on the spread of infectious disease. The Four Horsemen of the Apocalypse are War, Famine, Pestilence, and Death. They work in concert: for example, one of the biggest challenges to polio eradication has been the interruption of vaccination campaigns by wars in Yemen and Afghanistan. Perhaps the most effective way to rid ourselves of the worst of infectious disease would be to achieve world peace and global prosperity. Sadly, it seems easier to control infectious disease by understanding its ecology and evolution.

By the time you are reading this chapter, it seems virtually inevitable that yet another infectious disease will be emerging. *Plus ça change...* Apologize politely to the masked stranger who comes to your party uninvited, and explain that s/he will have to indulge you in hand washing before visiting the buffet; regale your guests with True Stories of effective vaccination campaigns; and take any medications you are given as directed (but feel free to discuss them with your doctor!). We cannot hope to live in a world free of infectious disease, but if we act wisely, we can live better and more safely.

Further reading

Chapter 1: Infection is inevitable

Blake, John B. 1952. 'The Inoculation Controversy in Boston: 1721-1722'. *The New England Quarterly* 25 (4): 489–506.

Brault, Aaron C., Claire Y.-H. Huang, Stanley A. Langevin, Richard M. Kinney, Richard A. Bowen, Wanichaya N. Ramey, Nicholas A. Panella, Edward C. Holmes, Ann M. Powers, and Barry R. Miller. 2007. 'A Single Positively Selected West Nile Viral Mutation Confers Increased Virogenesis in American Crows'. *Nature Genetics* 39 (9): 1162–6.

Combes, Claude. 2005. *The Art of Being a Parasite*. Chicago: University of Chicago Press.

Conniff, Richard. 2023. *Ending Epidemics: A History of Escape from Contagion*. Cambridge, Massachusetts: The MIT Press.

Cressler, Clayton E., David V. McLeod, Carly Rozins, Josée Van Den Hoogen, and Troy Day. 2016. 'The Adaptive Evolution of Virulence: A Review of Theoretical Predictions and Empirical Tests'. *Parasitology* 143 (7) (June): 915–30. <https://doi.org/10.1017/S003118201500092X>.

Davies, Julian. 1995. 'Vicious Circles: Looking Back on Resistance Plasmids'. *Genetics* 139 (4): 1465.

D'Costa, Vanessa M., Christine E. King, Lindsay Kalan, Mariya Morar, Wilson W. L. Sung, Carsten Schwarz, Duane Froese, et al. 2011. 'Antibiotic Resistance Is Ancient'. *Nature* 477 (7365): 457–61.

Mackowiak, Philip A., and Paul S. Sehdev. 2002. 'The Origin of Quarantine'. *Clinical Infectious Diseases* 35 (9): 1071–2.

Miranda, Mary Elizabeth G., and Noel Lee J. Miranda. 2011. 'Reston
Ebolavirus in Humans and Animals in the Philippines: A Review'.
Journal of Infectious Diseases 204 (suppl. 3): S757–60.

Mühlemann, Barbara, Lasse Vinner, Ashot Margaryan, Helene
Wilhelmson, Constanza de la Fuente Castro, Morten E. Allentoft,
Peter de Barros Damgaard, et al. 2020. 'Diverse Variola Virus
(Smallpox) Strains Were Widespread in Northern Europe in the
Viking Age'. *Science* 369 (6502): eaaw8977.

Quammen, David. 2013. *Spillover: Animal Infections and the Next
Human Pandemic*. Illustrated edition. New York: W. W. Norton &
Company.

Stearns, Stephen C., and Jacob C. Koella. 2008. *Evolution in Health
and Disease*. 2nd edition. Oxford; New York: Oxford
University Press.

Chapter 2: Transmission at different scales

Adam, David. 2020. 'Special Report: The Simulations Driving the
World's Response to COVID-19'. *Nature* 580 (7803) (2 April):
316–18. <https://doi.org/10.1038/d41586-020-01003-6>.

Bernoulli, Daniel, and Sally Blower. 2004. 'An Attempt at a New
Analysis of the Mortality Caused by Smallpox and of the
Advantages of Inoculation to Prevent It'. *Reviews in Medical
Virology* 14 (5): 275–88.

Combes, Claude. 2004. *Parasitism: The Ecology and Evolution of
Intimate Interactions*. Translated by Isaure de Buron and
Vincent A. Connors. Chicago: University of Chicago Press.

Gladwell, Malcolm. 2002. 'Fred Soper and the Global Malaria
Eradication Programme'. *Journal of Public Health Policy* 23
(4): 479–97.

Keeling, Matthew James, and Pejman Rohani. 2008. *Modeling
Infectious Diseases in Humans and Animals*. Princeton: Princeton
University Press.

McGovern, B., E. Doyle, L. E. Fenelon, and S. F. FitzGerald. 2010.
'The Necktie As a Potential Vector of Infection: Are Doctors
Happy to Do Without?' *The Journal of Hospital Infection* 75
(2): 138–9.

Schneider, David S., and Janelle S. Ayres. 2008. 'Two Ways to Survive
Infection: What Resistance and Tolerance Can Teach Us about
Treating Infectious Diseases'. *Nature Reviews Immunology* 8 (11)
(November): 889–95. <https://doi.org/10.1038/nri2432>.

Thomas, Y., G. Vogel, W. Wunderli, P. Suter, M. Witschi, D. Koch, C. Tapparel, and L. Kaiser. 2008. 'Survival of Influenza Virus on Banknotes'. *Applied and Environmental Microbiology* 74 (10): 3002–7.

Chapter 3: Influenza

Adams, Patrick. 2012. 'The Influenza Enigma'. *Bulletin of the World Health Organization* 90 (4): 245.

Cohen, Jon, and David Malakoff. 2012. 'On Second Thought, Flu Papers Get Go-Ahead'. *Science* 336 (6077): 19–20.

Dushoff, Jonathan, Joshua B. Plotkin, Cécile Viboud, David J. D. Earn, and Lone Simonsen. 2006. 'Mortality Due to Influenza in the United States: An Annualized Regression Approach Using Multiple-Cause Mortality Data'. *American Journal of Epidemiology* 163 (2): 181–7.

Francis, Magen Ellen, Morgan Leslie King, and Alyson Ann Kelvin. 2019. 'Back to the Future for Influenza Preimmunity—Looking Back at Influenza Virus History to Infer the Outcome of Future Infections'. *Viruses* 11 (2): 122. <https://doi.org/10.3390/v11020122>.

Huang, Karen E., Marc Lipsitch, Jeffrey Shaman, and Edward Goldstein. 2014. 'The US 2009 A/H1N1 Influenza Epidemic: Quantifying the Impact of School Openings on the Reproductive Number'. *Epidemiology* (Cambridge, Mass.) 25 (2): 203–6. <https://doi.org/10.1097/EDE.0000000000000055>.

Lipsitch, M., and C. Viboud. 2009. 'Influenza Seasonality: Lifting the Fog'. *Proceedings of the National Academy of Sciences* 106 (10): 3645–6.

Loeb, M., M. L. Russell, L. Moss, et al. 2010. 'Effect of Influenza Vaccination of Children on Infection Rates in Hutterite Communities: A Randomized Trial'. *JAMA* 303 (10): 943–50.

Nobusawa, Eri, and Katsuhiko Sato. 2006. 'Comparison of the Mutation Rates of Human Influenza A and B Viruses'. *Journal of Virology* 80 (7): 3675–8.

Roach, Jared C., Gustavo Glusman, Arian F. A. Smit, Chad D. Huff, Robert Hubley, Paul T. Shannon, Lee Rowen, et al. 2010. 'Analysis of Genetic Inheritance in a Family Quartet by Whole-Genome Sequencing'. *Science* 328 (5978): 636–9.

Simonsen, Lone, Peter Spreeuwenberg, Roger Lustig, Robert J. Taylor, Douglas M. Fleming, Madelon Kroneman, Maria D. Van Kerkhove, Anthony W. Mounts, W. John Paget, and the GLaMOR

Further reading

127

Collaborating Teams. 2013. 'Global Mortality Estimates for the 2009 Influenza Pandemic from the GLaMOR Project: A Modeling Study'. *PLoS Med* 10 (11): e1001558.

Tamerius, James, Martha I. Nelson, Steven Z. Zhou, Cécile Viboud, Mark A. Miller, and Wladimir J. Alonso. 2011. 'Global Influenza Seasonality: Reconciling Patterns across Temperate and Tropical Regions'. *Environmental Health Perspectives* 119 (4): 439–45.

Valkenburg, Sophie A., and Leo L. M. Poon. 2022. 'Exploring the Landscape of Immune Responses to Influenza Infection and Vaccination'. *Nature Medicine* 28 (2): 239–40. <https://doi.org/10.1038/s41591-021-01656-4>.

Viboud, Cécile, Katelyn Gostic, Martha I. Nelson, Graeme E. Price, Amanda Perofsky, Kaiyuan Sun, Nídia Sequeira Trovão, Benjamin J. Cowling, Suzanne L. Epstein, and David J. Spiro. 2020. 'Beyond Clinical Trials: Evolutionary and Epidemiological Considerations for Development of a Universal Influenza Vaccine'. *PLOS Pathogens* 16 (9): e1008583. <https://doi.org/10.1371/journal.ppat.1008583>.

Chapter 4: HIV/AIDS

Bouwman, Abigail, Natallia Shved, Gülfirde Akgül, Frank Rühli, and Christina Warinner. 2017. 'Ancient DNA Investigation of a Medieval German Cemetery Confirms Long-Term Stability of CCR5-Δ32 Allele Frequencies in Central Europe'. *Human Biology* 89 (2): 119–24. <https://doi.org/10.13110/humanbiology.89.2.02>.

Faria, Nuno R., Andrew Rambaut, Marc A. Suchard, Guy Baele, Trevor Bedford, Melissa J. Ward, Andrew J. Tatem, et al. 2014. 'The Early Spread and Epidemic Ignition of HIV-1 in Human Populations'. *Science* 346 (6205): 56–61.

Frank, Steven A. 2002. *Immunology and Evolution of Infectious Disease*. Princeton: Princeton University Press.

Fraser, C., K. Lythgoe, G. E. Leventhal, G. Shirreff, T. D. Hollingsworth, S. Alizon, and S. Bonhoeffer. 2014. 'Virulence and Pathogenesis of HIV-1 Infection: An Evolutionary Perspective'. *Science* 343 (6177).

Holmes, Edward C. 2009. *The Evolution and Emergence of RNA Viruses*. Oxford: Oxford University Press.

Hummel, S., D. Schmidt, B. Kremeyer, B. Herrmann, and M. Oppermann. 2005. 'Detection of the CCR5-Δ32 HIV

Resistance Gene in Bronze Age Skeletons'. *Genes & Immunity* 6 (4): 371–4. <https://doi.org/10.1038/sj.gene.6364172>.

McLaren, Paul J., and Mary Carrington. 2015. 'The Impact of Host Genetic Variation on Infection with HIV-1'. *Nature Immunology* 16 (6): 577–83. <https://doi.org/10.1038/ni.3147>.

Mild, Mattias, Rebecca R. Gray, Anders Kvist, Philippe Lemey, Maureen M. Goodenow, Eva Maria Fenyö, Jan Albert, Marco Salemi, Joakim Esbjörnsson, and Patrik Medstrand. 2013. 'High Intrapatient HIV-1 Evolutionary Rate Is Associated with CCR5-to-CXCR4 Coreceptor Switch'. *Infection, Genetics and Evolution* 19 (October): 369–77.

Müller, Viktor, and Sebastian Bonhoeffer. 2008. 'Intra-Host Dynamics and Evolution of HIV Infection'. In *Origin and Evolution of Viruses*, 2nd edition, edited by Esteban Domingo, Colin R. Parrish, and John J. Holland, 279–301. London: Academic Press.

Ndung'u, Thumbi, Joseph M. McCune, and Steven G. Deeks. 2019. 'Why and Where an HIV Cure Is Needed and How It Might Be Achieved'. *Nature* 576 (7787): 397–405. <https://doi.org/10.1038/s41586-019-1841-8>.

Novembre, John, and Eunjung Han. 2012. 'Human Population Structure and the Adaptive Response to Pathogen-Induced Selection Pressures'. *Philosophical Transactions of the Royal Society B: Biological Sciences* 367 (1590): 878–86.

Pépin, Jacques. 2021. *The Origins of AIDS*. 2nd edition. Cambridge: Cambridge University Press.

Quammen, David. 2015. *The Chimp and the River: How AIDS Emerged from an African Forest*. 1st edition. New York: W. W. Norton & Company.

Chapter 5: Cholera

Colwell, Rita R. 1996. 'Global Climate and Infectious Disease: The Cholera Paradigm'. *Science*, New Series 274 (5295): 2025–31.

Deen, Jacqueline, Martin A. Mengel, and John D. Clemens. 2020. 'Epidemiology of Cholera'. *Vaccine* 38 (S1): A31–40.

Gordillo Altamirano, Fernando L. and Jeremy J. Barr. 2019. 'Phage Therapy in the Postantibiotic Era'. *Clinical Microbiology Reviews* 32 (2): e00066–18.

Kitaoka, Maya, Sarah T. Miyata, Daniel Unterweger, and Stefan Pukatzki. 2011. 'Antibiotic Resistance Mechanisms of *Vibrio cholerae*'. *Journal of Medical Microbiology* 60 (4): 397–407.

Mogasale, Vittay, Vijayalaxmi J. Mogasale, and Amber Hsaio. 2020. 'Economic Burden of Cholera in Asia'. *Vaccine* 38 (S1): A160–6.

Morris, J. Glenn. 2011. 'Cholera: Modern Pandemic Disease of Ancient Lineage'. *Emerging Infectious Diseases* 17 (11): 2099–104.

Nelson, Eric J., Ashrafuzzaman Chowdhury, James Flynn, Stefan Schild, Lori Bourassa, Yue Shao, Regina C. LaRocque, Stephen B. Calderwood, Firdausi Qadri, and Andrew Camilli. 2008. 'Transmission of *Vibrio cholerae* Is Antagonized by Lytic Phage and Entry into the Aquatic Environment'. *PLoS Pathog* 4 (10): e1000187.

Nick, Jerry A. et al. 2022. Host and Pathogen Response to Bacteriophage Engineered against *Mycobacterium abscessus* Lung Infection'. *Cell* 185: 1860–74.

Zuger, Abigail. 2011. 'Small Fixes: Folding Saris to Filter Cholera-Contaminated Water'. *The New York Times*, 26 September. <http://www.nytimes.com/2011/09/27/health/27sari.html>.

Chapter 6: Malaria

Centers for Disease Control. 2019. 'Malaria Worldwide—How Can Malaria Cases and Deaths Be Reduced?—Indoor Residual Spraying'. <https://www.cdc.gov/malaria/malaria_worldwide/reduction/irs.html>.

Faust, E. C. 1951. 'The History of Malaria in the United States'. *American Scientist* 39 (1): 121–30.

Gladwell, Malcolm. 2002. 'Fred Soper and the Global Malaria Eradication Programme'. *Journal of Public Health Policy* 23 (4): 479–97.

Harper, Kristin N., and George J. Armelagos. 2013. 'Genomics, the Origins of Agriculture, and Our Changing Microbe-Scape: Time to Revisit Some Old Tales and Tell Some New Ones'. *American Journal of Physical Anthropology* 152 (S57): 135–52. <https://doi.org/10.1002/ajpa.22396>.

Hedrick, Philip W. 2012. 'Resistance to Malaria in Humans: The Impact of Strong, Recent Selection'. *Malaria Journal* 11 (October): 349.

Kantele, Anu, and T. Sakari Jokiranta. 2011. 'Review of Cases with the Emerging Fifth Human Malaria Parasite, *Plasmodium knowlesi*'. *Clinical Infectious Diseases* 52 (11): 1356–62.

Klayman, D. L. 1985. 'Qinghaosu (artemisinin): An Antimalarial Drug from China'. *Science* 228 (4703): 1049–55.

Lalremruata, Albert, Markus Ball, Raffaella Bianucci, Beatrix Welte, Andreas G. Nerlich, Jürgen F. J. Kun, and Carsten M. Pusch. 2013. 'Molecular Identification of Falciparum Malaria and Human Tuberculosis Co-Infections in Mummies from the Fayum Depression (Lower Egypt)'. *PLoS ONE* 8 (4): e60307.

Loy, Dorothy E., Weimin Liu, Yingying Li, Gerald H. Learn, Lindsey J. Plenderleith, Sesh A. Sundararaman, Paul M. Sharp, and Beatrice H. Hahn. 2017. 'Out of Africa: Origins and Evolution of the Human Malaria Parasites Plasmodium Falciparum and Plasmodium Vivax'. *International Journal for Parasitology*, Molecular Approaches to Malaria 2016 (MAM2016), 47 (2): 87–97. <https://doi.org/10.1016/j.ijpara.2016.05.008>.

MacDonald, Andrew J., and Erin A. Mordecai. 2019. 'Amazon Deforestation Drives Malaria Transmission, and Malaria Burden Reduces Forest Clearing'. *Proceedings of the National Academy of Sciences* 116 (44): 22212–18.

Mayer, Francois. 2020. 'The Quest for a Vaccine against Malaria'. *Nature Research*, September. <https://www.nature.com/articles/d42859-020-00021-8>.

Nosten, François H. 2014. 'How to Beat Malaria, Once and for All'. *The New York Times*, 7 June. <http://www.nytimes.com/2014/06/08/opinion/sunday/how-to-beat-malaria-once-and-for-all.html>.

Outlaw, Diana C., and Robert E. Ricklefs. 2011. 'Rerooting the Evolutionary Tree of Malaria Parasites'. *Proceedings of the National Academy of Sciences* 108 (32): 13183–7.

Piel, Frédéric B., Anand P. Patil, Rosalind E. Howes, Oscar A. Nyangiri, Peter W. Gething, Thomas N. Williams, David J. Weatherall, and Simon I. Hay. 2010. 'Global Distribution of the Sickle Cell Gene and Geographical Confirmation of the Malaria Hypothesis'. *Nature Communications* 1: 104.

Poinar, G., and S. R. Telford. 2005. '*Paleohaemoproteus burmacis* Gen. N., Sp. N. (Haemospororida: Plasmodiidae) from an Early Cretaceous Biting Midge (Diptera: Ceratopogonidae)'. *Parasitology* 131 (01): 79–84.

Rich, Stephen M., Monica C. Licht, Richard R. Hudson, and Francisco J. Ayala. 1998. 'Malaria's Eve: Evidence of a Recent Population Bottleneck throughout the World Populations of *Plasmodium falciparum*'. *Proceedings of the National Academy of Sciences* 95 (8): 4425–30.

Rich, Stephen M., and Guang Xu. 2011. 'Resolving the Phylogeny of
  Malaria Parasites'. *Proceedings of the National Academy of Sciences*
  108 (32): 12973–4.

Sarma, Nayantara, Edith Patouillard, Richard E. Cibulskis, and
  Jean-Louis Arcand. 2019. 'The Economic Burden of Malaria:
  Revisiting the Evidence'. *The American Journal of Tropical
  Medicine and Hygiene* 101 (6): 1405–15. <https://doi.org/10.4269/
  ajtmh.19-0386>.

Tishkoff, Sarah A., Robert Varkonyi, Nelie Cahinhinan, Salem Abbes,
  George Argyropoulos, Giovanni Destro-Bisol, Anthi Drousiotou,
  et al. 2001. 'Haplotype Diversity and Linkage Disequilibrium at
  Human G6PD: Recent Origin of Alleles that Confer Malarial
  Resistance'. *Science* 293 (5529): 455–62.

Webb, James L. A. 2009. 'The Long Shadow of Malaria Interventions
  in Tropical Africa'. *The Lancet* 374 (9705): 1883–4.

World Health Organization. 2021. 'Vaccine Efficacy, Effectiveness and
  Protection.' 14 July. <https://www.who.int/news-room/feature-
  stories/detail/vaccine-efficacy-effectiveness-and-protection>.

Chapter 7: Amphibian chytrid fungus

Basanta, M. Delia, Allison Q. Byrne, Erica Bree Rosenblum, Jonah
  Piovia-Scott, and Gabriela Parra-Olea. 2021. 'Early Presence of
  Batrachochytrium Dendrobatidis in Mexico with a Contemporary
  Dominance of the Global Panzootic Lineage'. *Molecular Ecology*
  30 (2): 424–37. <https://doi.org/10.1111/mec.15733>.

Blaustein, Andrew R., John M. Romansic, Erin A. Scheessele,
  Barbara A. Han, Allan P. Pessier, and Joyce E. Longcore. 2005.
  'Interspecific Variation in Susceptibility of Frog Tadpoles to the
  Pathogenic Fungus *Batrachochytrium dendrobatidis*'.
  Conservation Biology 19 (5): 1460–8.

Briggs, Cheryl J., Vance T. Vredenburg, Roland A. Knapp, and
  Lara J. Rachowicz. 2005. 'Investigating the Population-Level
  Effects of Chytridiomycosis: An Emerging Infectious Disease of
  Amphibians'. *Ecology* 86 (12): 3149–59.

Burke, Katie L. 2013. 'Probiotics for Frogs'. *American Scientist* 100
  (3): 190.

Fisher, Matthew C., Trenton W. J. Garner, and Susan F. Walker. 2009.
  'Global Emergence of *Batrachochytrium dendrobatidis* and
  Amphibian Chytridiomycosis in Space, Time, and Host'. *Annual
  Review of Microbiology* 63 (1): 291–310.

Fisher, Matthew C., and Trenton W. J. Garner. 2020. 'Chytrid Fungi and
    Global Amphibian Declines'. *Nature Reviews Microbiology* 18 (6)
    (June): 332–43. <https://doi.org/10.1038/s41579-020-0335-x>.

Gervasi, Stephanie S., Jenny Urbina, Jessica Hua, Tara Chestnut,
    Rick A. Relyea, and Andrew R. Blaustein. 2013. 'Experimental
    Evidence for American Bullfrog (*Lithobates catesbeianus*)
    Susceptibility to Chytrid Fungus (*Batrachochytrium dendrobatidis*)'.
    EcoHealth 10 (2): 166–71.

Kilpatrick, A. Marm, Cheryl J. Briggs, and Peter Daszak. 2010. 'The
    Ecology and Impact of Chytridiomycosis: An Emerging Disease of
    Amphibians'. *Trends in Ecology & Evolution* 25 (2): 109–18.

Lips, Karen R., Jay Diffendorfer, Joseph R. Mendelson III, and
    Michael W. Sears. 2008. 'Riding the Wave: Reconciling the Roles
    of Disease and Climate Change in Amphibian Declines'. *PLoS Biol*
    6 (3): e72.

McMahon, Taegan A., Laura A. Brannelly, Matthew W. H. Chatfield,
    Pieter T. J. Johnson, Maxwell B. Joseph, Valerie J. McKenzie,
    Corinne L. Richards-Zawacki, Matthew D. Venesky, and
    Jason R. Rohr. 2013. 'Chytrid Fungus *Batrachochytrium
    dendrobatidis* Has Nonamphibian Hosts and Releases Chemicals
    that Cause Pathology in the Absence of Infection'. *Proceedings of
    the National Academy of Sciences* 110 (1): 210–15.

McMahon, Taegan A., Brittany F. Sears, Matthew D. Venesky,
    Scott M. Bessler, Jenise M. Brown, Kaitlin Deutsch,
    Neal T. Halstead, et al. 2014. 'Amphibians Acquire Resistance to
    Live and Dead Fungus Overcoming Fungal Immunosuppression'.
    Nature 511 (7508): 224–7.

Martel, An, Annemarieke Spitzen-van der Sluijs, Mark Blooi, Wim
    Bert, Richard Ducatelle, Matthew C. Fisher, Antonius Woeltjes,
    et al. 2013. '*Batrachochytrium salamandrivorans* Sp. Nov. Causes
    Lethal Chytridiomycosis in Amphibians'. *Proceedings of the
    National Academy of Sciences* 110 (38): 15325–9.

Rachowicz, Lara J., Roland A. Knapp, Jess A. T. Morgan,
    Mary J. Stice, Vance T. Vredenburg, John M. Parker, and
    Cheryl J. Briggs. 2006. 'Emerging Infectious Disease as a
    Proximate Cause of Amphibian Mass Mortality'. *Ecology* 87 (7):
    1671–83.

Rodriguez, D., C. G. Becker, N. C. Pupin, C. F. B. Haddad, and
    K. R. Zamudio. 2014. 'Long-Term Endemism of Two Highly
    Divergent Lineages of the Amphibian-Killing Fungus in the
    Atlantic Forest of Brazil'. *Molecular Ecology* 23 (4): 774–87.

Rohr, Jason R., and Thomas R. Raffel. 2010. 'Linking Global Climate and Temperature Variability to Widespread Amphibian Declines Putatively Caused by Disease'. *Proceedings of the National Academy of Sciences* 107 (18): 8269–74.

Rosenblum, Erica Bree, Timothy Y. James, Kelly R. Zamudio, Thomas J. Poorten, Dan Ilut, David Rodriguez, Jonathan M. Eastman, et al. 2013. 'Complex History of the Amphibian-Killing Chytrid Fungus Revealed with Genome Resequencing Data'. *Proceedings of the National Academy of Sciences* 110 (23): 9385–90.

Chapter 8: SARS-CoV-2/COVID-19

Billah, Md Arif, Md Mamun Miah, and Md Nuruzzaman Khan. 2020. 'Reproductive Number of Coronavirus: A Systematic Review and Meta-Analysis Based on Global Level Evidence'. *PLOS ONE* 15 (11) (11 November): e0242128. <https://doi.org/10.1371/journal.pone.0242128>.

Holmes, Edward C., Stephen A. Goldstein, Angela L. Rasmussen, David L. Robertson, Alexander Crits-Christoph, Joel O. Wertheim, Simon J. Anthony, et al. 2021. 'The Origins of SARS-CoV-2: A Critical Review'. *Cell* 184 (19) (16 September): 4848–56. <https://doi.org/10.1016/j.cell.2021.08.017>.

Pekar, Jonathan E., Andrew Magee, Edyth Parker, Niema Moshiri, Katherine Izhikevich, Jennifer L. Havens, Karthik Gangavarapu, et al. 2022. 'The Molecular Epidemiology of Multiple Zoonotic Origins of SARS-CoV-2'. *Science* (July). <https://doi.org/10.1126/science.abp8337>.

Quammen, David. 2022. *Breathless: The Scientific Race to Defeat a Deadly Virus*. New York: Simon & Schuster.

Rosa, Sara Sousa, Duarte M. F. Prazeres, Ana M. Azevedo, and Marco P. C. Marques. 2021. 'mRNA Vaccines Manufacturing: Challenges and Bottlenecks'. *Vaccine* 39 (16): 2190–200. <https://doi.org/10.1016/j.vaccine.2021.03.038>.

Segreto, Rossana, Yuri Deigin, Kevin McCairn, Alejandro Sousa, Dan Sirotkin, Karl Sirotkin, Jonathan J. Couey, Adrian Jones, and Daoyu Zhang. 2021. 'Should We Discount the Laboratory Origin of COVID-19?' *Environmental Chemistry Letters* 19 (4) (1 August): 2743–57. <https://doi.org/10.1007/s10311-021-01211-0>.

Wang, Haidong, Katherine R. Paulson, Spencer A. Pease, Stefanie Watson, Haley Comfort, Peng Zheng, Aleksandr Y. Aravkin, et al. 2022. 'Estimating Excess Mortality Due to the COVID-19 Pandemic: A Systematic Analysis of COVID-19-Related Mortality, 2020–21'. *The Lancet* 399 (10334): 1513–36. <https://doi.org/10.1016/S0140-6736(21)02796-3>.

Worobey, Michael, Joshua I. Levy, Lorena Malpica Serrano, Alexander Crits-Christoph, Jonathan E. Pekar, Stephen A. Goldstein, Angela L. Rasmussen, et al. 2022. 'The Hunan Seafood Wholesale Market in Wuhan Was the Early Epicenter of the COVID-19 Pandemic'. *Science* 377 (6609) (26 July): abp8715. <https://doi.org/10.1126/science.abp8715>.

Yong, Ed. 2022. 'The Pandemic's Legacy Is Already Clear'. *The Atlantic*, 30 September. <https://www.theatlantic.com/health/archive/2022/09/covid-pandemic-exposes-americas-failing-systems-future-epidemics/671608/>.

Chapter 9: Looking ahead

Alamo, Teodoro, Daniel G. Reina, Pablo Millán Gata, Victor M. Preciado, and Giulia Giordano. 2021. 'Data-Driven Methods for Present and Future Pandemics: Monitoring, Modelling and Managing'. *Annual Reviews in Control* 52: 448–64. <https://doi.org/10.1016/j.arcontrol.2021.05.003>.

Brown, P., R. G. Will, R. Bradley, D. M. Asher, and L. Detwiler. 2001. 'Bovine Spongiform Encephalopathy and Variant Creutzfeldt–Jakob Disease: Background, Evolution, and Current Concerns'. *Emerging Infectious Diseases* 7 (1): 6–16.

Dudas, G., Carvalho, L., Bedford, T. et al. 2017. 'Virus Genomes Reveal Factors that Spread and Sustained the Ebola Epidemic'. *Nature* 544: 309–15.

Fisman, David, Edwin Khoo, and Ashleigh Tuite. 2014. 'Early Epidemic Dynamics of the West African 2014 Ebola Outbreak: Estimates Derived with a Simple Two-Parameter Model'. *PLOS Currents Outbreaks*. 8 September. Edition 1.

Hadfield, James, et al. 2018. 'Nextstrain: Real-Time Tracking of Pathogen Evolution'. *Bioinformatics* 34 (23): 4121–3.

Hewlett, Barry, and Bonnie Hewlett. 2007. *Ebola, Culture and Politics: The Anthropology of an Emerging Disease*. Andover: Cengage Learning.

Lessler, Justin, Lelia H. Chaisson, Lauren M. Kucirka, Qifang Bi, Kyra Grantz, Henrik Salje, Andrea C. Carcelen, et al. 2016. 'Assessing the Global Threat from Zika Virus'. *Science* 353 (6300): aaf8160. <https://doi.org/10.1126/science.aaf8160>.

Reiter, Paul, Christopher J. Thomas, Peter M. Atkinson, Simon I. Hay, Sarah E. Randolph, David J. Rogers, G. Dennis Shanks, Robert W. Snow, and Andrew Spielman. 2004. 'Global Warming and Malaria: A Call for Accuracy'. *The Lancet Infectious Diseases* 4 (6): 323–4.

Suzuki, Y., and T. Gojobori. 1997. 'The Origin and Evolution of Ebola and Marburg Viruses'. *Molecular Biology and Evolution* 14 (8): 800–6.

Wang, Zengmiao, et al. 2022. 'The Relationship between Rising Temperatures and Malaria Incidence in Hainan, China, from 1984 to 2010: A Longitudinal Cohort Study'. *The Lancet: Planetary Health* 6 (4): E350–8.

Index

For the benefit of digital users, indexed terms that span two pages (e.g., 52–53) may, on occasion, appear on only one of those pages.

EPIDEMIOLOGY
A Very Short Introduction
Rodolfo Saracci

Epidemiology has had an impact on many areas of medicine;
and lung cancer, to the origin and spread of new epidemics.
and lung cancer, to the origin and spread of new epidemics.
However, it is often poorly understood, largely due to
misrepresentations in the media. In this *Very Short Introduction*
Rodolfo Saracci dispels some of the myths surrounding the
study of epidemiology. He provides a general explanation of
the principles behind clinical trials, and explains the nature of
basic statistics concerning disease. He also looks at the ethical
and political issues related to obtaining and using information
concerning patients, and trials involving placebos.